MW00667995

NASOLOGY:

OR,

HINTS TOWARDS A CLASSIFICATION OF NOSES.

BY EDEN WARWICK.

"'Mayhap there is more meant than is said in it,' quoth my father 'Learned men, brother Toby, don't write dialogues upon long Noses for nothing.'"—TRISTRAM SHANDY.

LONDON:
RICHARD BENTLEY, NEW BURLINGTON STREET.
1848.

PREFACE.

WITH regard to a Preface to his Book, an Author has to contend with three great, but unequal, difficulties. The first and greatest, is to persuade his Publisher to issue it without a Preface; the next, is to write one himself; and the third and least, is to get some one to write it for him. Now there is a wise old saw which says, "Of divers evils choose the least;" and as the learned Slawkenbergius (so says Tristram Shandy) has prefaced his FOLIO on Noses with a clause which exactly explains our own qualifications, and reasons, for writing on the same important subject, we invoke him to relieve us of the third difficulty: "'ever since I understood,' quoth Slawkenbergius, 'anything—or rather *what was what*—and could perceive that the point of long Noses had been too loosely handled by all who had gone before—have I, Slawkenbergius, felt a strong impulse, with a mighty and irresistible call within me, to gird up myself to this undertaking.'"

Now this is exactly our own case, and must, therefore, suffice for our Preface; nevertheless, we cannot flatter ourselves that our brief hints will be eulogized, like the gigantic folio of Hafen Slawkenbergius, as "an institute of all that is necessary to be known of Noses." It professes to be nothing more than an introduction to the subject of Nasology; written originally for the use of friends, and afterwards extended for publication. This will account for some discrepancies which may be perceptible in the style—discrepancies which it was thought best not to remove, as the additions were on subjects of a more grave and important character than the original sketch; and, therefore, the diversities of style appeared to be rather consistent and advantageous.

MAY 26, 1848.

CONTENTS.

CHAPTER VII.

CHAPTER VIII.

CHAPTER IX.

NASOLOGY.

CHAPTER I.

OF THE CLASSIFICATION OF NOSES.

It has not been hastily, nor until after long and careful observation, that the theory propounded in the following pages has been published; a theory which, at first sight, may appear to some wild and absurd, to others simply ridiculous, to others wicked and heretical,*

* It would be rather amusing, if it were not a melancholy sign of human perverseness, to sum up all the hypotheses which have been at their first promulgation pronounced impious and heretical. The denial of the approaching End of the World in any century after Christ; the Copernican System; Inoculation and Vaccination for the Small-pox; the change of the Style of the year; Geology, Phrenology, &c., &c., would be included in the list of *umquhile* heresies.

and to others fraught with social mischief and danger.

Nevertheless, we shall not begin by deprecating the ridicule, or the censure of any one. The only vindication which an author is entitled to offer, is that which his works themselves present. If his cause is a good one, it requires no apology; if it is a bad one, to vindicate it is either useless or baneful; useless, if it blinds no one to his errors; baneful, if it induces any one blindly to receive his brass for sterling gold.

The only circumstance which can attach any value to our observations is, that they are entirely original, and wholly unbiassed by the theories of any other writers on physiognomy. When we commenced observing Noses, we just knew that some few forms of the Nose had names given them, as the Roman, the Greek, &c.; but we regarded these as mere artistic definitions of form, and were wholly ignorant what mental characteristics had been ascribed to them. So far as this nomenclature went, it appeared best to adopt it, as affording well-known designations of Nasal profiles; and our investigations were, therefore, commenced by endeavouring to discover whether these forms of Nose characterized any, and what, mental properties. In order to do this with accuracy, it was absolutely necessary still to keep the mind unacquainted with the system of any other writers, if such there were, lest it should unconsciously imbibe preconceptions and hints which would render its independent researches open to the suspicion of bias. We felt that if the characteristics attributed by us to Noses, after long and extensive observation, corresponded with those of any other writer, a powerful corroboration of our views would thus be gained.

It may happen, therefore, and it is hoped it will be so, that we may sometimes appear to have plagiarized from other physiognomists, and to have adopted their views; but this correspondence must, nevertheless, be accepted as a further proof of the accuracy of our honest independent labours.

It was impossible, however, amidst much multifarious reading, to keep the mind, latterly, wholly ignorant that some mental characteristics had been ascribed to Noses; but into the nature of these we never inquired, nor are we now aware that anything has been done, beyond the throwing out of a few unconnected, unattested hints, towards a systematic deduction of mental qualifications from Nasal formation.

If it is improper to vindicate one's self, it might not seem altogether unfitting to vindicate one's subject from ridicule; and it might appear prudent, if not altogether necessary, to commence by vindicating the Nose from the charge of being too ridiculous an organ to be seriously discoursed upon. But this ridiculousness is mere prejudice; intrinsically one part of the face is as worthy as another, and we may feel assured that He who gave the *os sublime* to man, did not place, as its foremost and most prominent feature, a *ridiculous* appendage.

To come then at once to our subject. We have a belief, founded on long-continued, personal observation, that there is more in a Nose than most owners of that appendage are generally aware. We believe that, besides being an ornament to the face, or a convenient handle by which to grasp an impudent fellow, it is an important index to its owner's character; and that the accurate observation and minute comparison of an extensive collection of Noses of persons whose mental characteristics are known, justifies a Nasal Classification, and a deduction of some points of mental organization therefrom. It will not be contended that all the faculties and properties of mind are revealed by the Nose;—for instance, we can read nothing of Temper or the Passions from it.* Perhaps it rather reveals Power and Taste—Power or Energy to carry out Ideas, and the Taste or Inclination which dictates or guides them. As these will always very much form a man's outward character, the proposition which is sought to be established is this:—"THE NOSE IS AN IMPORTANT INDEX TO CHARACTER."

* We shall endeavour to speak of mind in popular phraseology, instead of in the obscure terms in which metaphysicians envelop their ignorance of mental phenomena.

It may be prudent to observe that we utterly repudiate the doctrine of the Phrenologists, that the form of the Body affects the manifestations, and even properties, of the Mind.

We contend that the Mind forms the Nose, and not the Nose the Mind. We have carefully endeavoured to avoid phraseology which should induce a supposition that we entertain the latter absurdity; but here enter this protest once for all, lest a want of precision in our language, or the obtuseness of critics should cause us to be charged with it.

And here we might descant, at considerable length, and with much show of learning, on the influence of the Mind over the Body. We might impugn the wisdom of those who, undertaking to cure either, have forgotten that they were so intimately united and mutually dependent, that they could not be treated separately with success. We might shew that the first step of the physician towards curing mental disorder, is to free the body from disease; and that of him who would cure the body, is, ofttimes, to apply his remedies to the derangement of the mind. But, though by so doing we might swell our pages and eke out an additional chapter—an important consideration if we were a mere book-maker—we shall not, as we have some qualms of conscience whether it would be quite germane to the matter in hand. It might not, however, be out of place to remind the reader that physiognomy, or the form which mind gives to the features, is universally recognised. A pleasant mouth, a merry eye, a sour visage, a stern aspect are some of the common phrases by which we daily acknowledge ourselves to be physiognomists; for by these expressions we mean, not that the mouth is pleasant or the visage sour, but that such is the mind which shines out from them. If it were the face alone which we thus intended, we should never trouble or concern ourselves about a human countenance, nor be attracted, nor repulsed by one, any more than if it were a carved head on a gothic waterspout, or a citizen's door-knocker. We all acknowledge the impression given by the mind to the mouth and the eyes because they express Temper and the Passions—those feelings which more immediately interest us in our mutual intercourse—and because they change with the feelings; now flashing with anger, or sparkling with pleasure, compressing with rage, or smiling with delight.

But because the Nose is uninfluenced by the feelings which agitate and vary the mind, and, is, therefore, immoveable and unvaried, no one will hear the theory of Nasology broached without incredibility and risibility. Because the Nose is subject only to those faculties of mind which are permanent and unfluctuating; and is, therefore, likewise permanent and unfluctuating in its form, men have paid no attention to its indications, and will, accordingly, abuse as an empiric and dotard the first Nasologist. But, is there, *à priori*, any thing so unreasonable in attributing mental characteristics to the Nose, when we all daily read each other's minds in the Nose's next door neighbours, the eyes and mouth? Is not the *à priori* inference entirely in favour of a negative reply? And that, *à posteriori*, it may confidently be replied to in the negative will, it is hoped, presently appear.

There is here room for another long disquisition to point out the advantages of Nasology. How that the permanency and immobility of the Nose forbid hypocrisy to mould it to any artificial feelings, as the eyes and the mouth may be. And how this immobility, together with its prominency and incapability of being concealed, like bad phrenological bumps, render it a sure guide to some parts of our fellow-creature's mental organization. But it would be premature to do this before proving somewhat of the truth of Nasology; and when that is done, no one will deny that it has its uses, though it may be disputed what those are.

Nevertheless, we must earnestly protest against the fallacy of attempting to judge what any person *is* from his Nose; we can only judge of natural tendency and capacity—education and external circumstances of a thousand different kinds, may have swerved the mind from its original tendency, or prevented the development of inherent faculties. It is in this unfair and uncharitable asserting dogmatically the disposition and character, vices and virtues of a man, that phrenologists so greatly err; whereas they ought to confine their inferences from external development of organs, to capacity and tendency only.

The impossibility of giving such numerous pictorial illustrations as the subject properly demands, will confine the examples adduced to those only of which portraits are well known and easily accessible. If, therefore, the proofs are thought insufficient in number, it must be attributed to this circumstance alone. It would have been easy to have swelled them by a number of names, the right of which to be included in the lists the majority of persons would have been unable to verify. Nevertheless, the examples will be found much more numerous and more easily verifiable than those which have been deemed sufficient to establish Phrenology as an hypothesis, if not as a science; and, had we, like the principal expounder of

Phrenology,* dragged in as "proofs" nameless gentlemen of our acquaintance, we might have still further swelled the lists of examples. But it seemed to our humble judgment, to be demanding more from the reader's good nature than would be compatible with sound criticism, to ask him to accept such unsupported *dicta* as *proofs*. Of course, very many of the examples by which our own mind has been satisfied have been drawn from personal observation, among friends and acquaintance; and not only have these been the most numerous proofs, but also by far the most satisfactory, as they afforded the most exact and undeniable profiles, and the most noticeable mental characteristics. The slightest incorrectness in the artist, may render useless a pictorial example; but, when we are looking upon the original himself, there can be no mistake. A thousand minutiæ of character may escape a biographer, which appear plainly in the man himself.

Nevertheless, we felt so strongly how unfitting

* See Combe's Phrenology; *passim.*

it would be to offer such mere personal observations as *proofs*, that we have carefully refrained from admitting any example which is not open to the observation of almost every one.

This is a drawback which we feel greatly; it reduces our instances to a hundredth part of those which might be adduced; but we must submit to it, only asking of the reader's generosity to take it into account. Another favour which we beg is, that the reader will suspend his judgment until the subject is concluded, and he has the whole system with all its proofs before him.

We scruple not to admit, that at present the system is incomplete. We rather court inquiry, and solicit additional facts, than peremptorily dogmatize on conclusions drawn from our own limited—though extensive—number of observations. But it is so much the fashion for every wild theorist to dogmatize on his theory, and insist upon it, *per fas et nefas*, as perfect, unassailable, and complete, that it is almost deemed reprehensible to suggest a notion for the consideration of the world, or to propound anything which the author is modest enough to admit is improvable. Such, however, was not the manner of the true philosophers of former days. If Copernicus had delayed propounding the system of the universe which bears his name, until he could explain by it all the planetary and sidereal motions, it might have slumbered unknown for another century or two, and so we should not yet have arrived at our present enlarged understanding of it. If Bacon had waited for a complete Natural History, ere he published his Novum Organum, we might still have been groping after the Sciences with the dark-lanthorn of Aristotle and the schools. If Newton had withheld his theory of Light until he could burn a diamond, our knowledge of the nature of Light might still be in its infancy.

These examples must furnish an apology for submitting for candid consideration and further development, a theory which we believe to be well founded, but which is capable of improvement and extension.

Subject to the foregoing remarks, the following Physical Classification of Noses is submitted, as being, in part, well-known and long-established, because well-defined and clearly marked:—

Class I. THE ROMAN, or Aquiline Nose.
„ II. THE GREEK, or Straight Nose.
„ III. THE COGITATIVE, or Wide-nostrilled Nose.
„ IV. THE JEWISH, or Hawk Nose.
„ V. THE SNUB Nose, and
„ VI. THE CELESTIAL, or Turn-up Nose.

Between these there are infinite crosses and intermixtures which will at first embarrass the student, but which after a little practice, he will be able to distinguish with tolerable precision. A compound of different Noses will of course indicate a compound character; and it is only in the rather rare instance of a perfect Nose of any of the classes that we find a character correspondingly strongly developed. We shall endeavour to support each part of the hypothesis by well-defined and striking instances, selecting the most decided and perfect noses of each class, and at the same time the most peculiar and decided characters.

Class I. The Roman, or Aquiline Nose, is rather convex, but undulating as its name aquiline imports. It is usually rugose and coarse; but when otherwise it approaches the Greek nose, and the character is materially altered.

It indicates great Decision, considerable Energy, Firmness, Absence of refinement, and Disregard for the *bienséances* of life.

Class II. The Greek, or Straight Nose, is perfectly straight; any deviation from the right line must be strictly noticed. If the deviation tend to convexity, it approaches the Roman Nose, and the character is improved by an accession of energy; on the other hand, when the deviation is towards concavity, it partakes of the "Celestial," and the character is weakened. It should be fine and well-chiselled, but not sharp.

It indicates Refinement of character, Love for the fine arts and *belles-lettres*, Astuteness, craft and a preference for indirect, rather than direct action. Its owner is not without some energy in pursuit of that which is agreeable to his tastes; but, unlike the owner of the Roman Nose, he cannot exert himself in *opposition* to his tastes. When associated with the Roman Nose, and distended slightly at the end by the Cogitative, it indicates the most useful and intellectual of characters; and is the highest and most beautiful form which the organ can assume.*

---

* The Platonic theory that beauty of form generally indicates beauty of mind, is finely condensed by Spenser into a single line:

"All that is good is beautiful and fair."

A HYMN OF HEAVENLY BEAUTY.

And again:

. "All that fair is, is by nature good;
That is a sign to know the gentle blood."

IBID.

Wordsworth would also appear to be a Platonist:

"For passions link'd to forms so fair
And stately, needs must have their share
Of noble sentiment."

RUTH.

Class III. The Cogitative, or Wide-nostrilled Nose, is, as its secondary name imports, wide at the end, thick and broad; not clubbed, but *gradually* widening from below the bridge. The other noses are seen in profile, but this in full face.

It indicates a Cogitative mind, having strong powers of Thought, and given to close and serious Meditation. Its indications are of course much dependent on the form of the Nose in profile, which decides the turn the cogitative power will take. Of course, it never occurs alone; and is usually associated with Classes I and II, rarely with IV, still more seldom with V and VI.* The entire absence of it produces the "sharp" nose, which is not classified, as sharpness is only a negative quality, being the defect of breadth,† and, therefore, indicates defect of cogitative power.

* A Nose should never be judged of in profile only; but should be examined also in front to see whether it partakes of Class III.

† Thus Phrenologists rightly urge that negative

CLASS IV. THE JEWISH, or Hawk Nose, is very convex, and preserves its convexity like a bow, throughout the whole length from the eyes to the tip. It is thin and sharp.

It indicates considerable Shrewdness in worldly matters; a deep Insight into character, and facility of turning that insight to profitable account.

CLASSES V AND VI. THE SNUB Nose, and the Turn-up, *poeticè* CELESTIAL Nose. The form of the former is sufficiently indicated by its name. The latter is distinguished by its presenting a continuous concavity from the eyes to the tip. It is converse in shape to the Jewish nose.

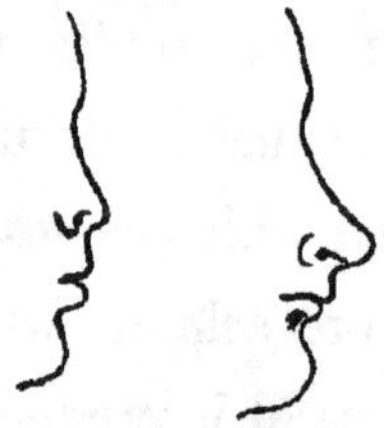

N.B. The Celestial must not be confounded

qualities require no organ. Hate, is only the absence of Benevolence; dislike to children, a defective Philoprogenitiveness.

with a Nose, which, belonging to one of the other classes in the upper part, terminates in a slight distension of the tip; for this, so far from prejudicing the character, rather adds to its warmth and activity.

We associate the Snub and the Celestial in nearly the same category, as they both indicate natural weakness, mean, disagreeable disposition, with petty insolence, and divers other characteristics of conscious weakness, which strongly assimilate them (indeed, a true Celestial Nose is only a Snub turned up); while their general poverty of distinctive character, makes it almost impossible to distinguish them. Nevertheless there is a difference between their indications; arising, however, rather from difference of intensity than of character. The Celestial is, by virtue of its greater length, decidedly preferable to the Snub; as it has all the above unfortunate propensities in a much less degree, and is not without some share of small shrewdness, and fox-like common sense; on which, however, it is apt to presume, and is, therefore, a more impudent Nose than the Snub.

The following subordinate rules are applicable to all Noses, and must be attended to before forming a judgment on any Nose.

1. The character of a Nose is weakened in intensity by forming too great, or too small an angle with the general profile of the face. This angle, if as great as 40^0 is not good, anything beyond that is bad; about 30^0 is best. Angles:

45°. 40°. 30°. less than 25°,

become a snub.

2. Attention should be paid to the angle which the basal line of the Nose forms with the upper lip. This angle affects intensity, and also temperament. If it is an obtuse angle, as thus  , the consequent abbreviation of the Nose (for a long Nose has always more Power than a short one) weakens the character, but the temperament is cheerful, gay and lively; if on the other hand the angle is acute, as thus  , the elongation of the Nose

adds much to the intensity of the character indicated by the profile; but the disposition is generally melancholy, and, if a very acute angle, desponding and fond of gloomy thoughts. Dante, Fox (the Martyrologist), John Knox, Calvin, E. Spenser, and George Herbert, are illustrations of the melancholy Nose.

DANTE.

CHAPTER II.

OF THE ROMAN NOSE.

Class I.—The Roman, or Aquiline Nose, is rather convex, but undulating, as its name aquiline imports. It is usually rugose and coarse; but when otherwise it approaches the Greek Nose, and the character is materially altered.

It indicates great Decision, considerable Energy, Firmness, Absence of refinement, and Disregard for the *bienséances* of life.

Numerous portraits, both in marble and on coins, demonstrate that this Nose was very frequent among the Romans, and peculiarly characteristic of that nation. Hence its name. The persevering energy, stern determination and unflinching firmness of the conquerors of the world; their rough, unrefined character, which, notwithstanding the example of Greece, never acquired the polish of that country, all indicate the accuracy of the mental habit attributed to the owner of this Nose.

Sufficient stress has never been laid by historians on national characteristics. The peculiar psychonomy of nations is an element which is never taken into account, when the historical critic endeavours to elucidate the causes and consequences of events. He judges of all nations by the standard of his own, regardless of age, climate, physiognomy and psychonomy. This is as absurd as the fashion the Greeks had of deducing foreign names and titles from the Greek, a practice which Cicero wittily ridicules. In this ridicule we willingly join; yet we are equally open to it, when we interpret the action of foreign nations by our own national standard.

It was the psychonomic difference between the Romans and the Greeks, which prevented the former from benefiting so efficiently from the lessons in art and philosophy of the latter as they would have done, had their minds been congenial.

The refinement which Rome received from Greece, was converted in the transfer into a refinement of coarse sensual luxury. Rome after the conquest of Greece filled its forums and halls with Greek workmanship, and its schools with Greek learning; nevertheless Roman mind advanced not one step beyond its original coarseness.

At the period when Rome possessed itself by conquest of the principal works of Grecian art, her citizens only regarded them as household furniture of but little value. Polybius narrates that, after the siege of Corinth, he saw some Roman soldiers playing at dice upon a picture of Bacchus by Aristides; a picture esteemed one of the finest in the world. When King Attalus offered 600,000 sesterces, (£4,845 15*s.*) for this picture, Mummius, the Roman Consul, thinking there must be some magic property in it, to make it worth such an enormous sum, refused to sell it, and hung it up in the Temple of Ceres at Rome. So little were the Romans conscious of the real value of the treasures of Greek art, that Mummius covenanted with the masters of the ships, hired to convey the spoils of Corinth to Rome, that if any of the exquisite paintings and statuary should be lost, *they should replace them with new ones!**

It is not surprising, therefore, that Rome, although possessed of infinitely greater wealth, a larger population, and the splendid examples of Greece, not only produced no artist of merit, but receded far from the high standard which Greece, notwithstanding its internal divisions, its comparative poverty, small extent, and unassisted genius, had established. There is no way of accounting for these facts, but by the difference in their psychonomy. The genius of Rome was of a very different nature from that of Greece, and was incompetent to advance the great work which the latter had commenced.

This is an example which, with numerous others that occur in the world's history, might teach those who, in modern phrase, assert that

* Hooke's Rom. Hist. B. vi. c. i.

the uniform order of the world is *progress*, that retrogression has ofttimes been the apparent order, and that it is a foolish short-sightedness to judge of the order of the world from a few hundred years in its history. The Greek who remembered the magnificent works of his country, and looked upon the degenerate splendour of Rome, no doubt equally dogmatically, asserted that the world was in its dotage, that it had retrograded, and would never be regenerated.

But the modern dogmatist tries to take his case out of the argument, by pretending that Christianity will protect the world from again retrograding. This is the mere pride of the Pharisee, who flatters himself that he is not as other men are, that his Christianity is too pure to fall, and his knowledge too vast to be blasted. Or else he forgets that the pure Christianity of the first disciples and martyrs failed to preserve succeeding generations from the inroads of sin and darkness more overwhelming than had ever blackened the face of Europe since the commencement of the historical period. The dogmatist of those days sighed over the world's degeneracy, and saw not through the surrounding gloom, an hopeful gleam of light; just as the modern dogmatist rejoices over the world's advance, without perceiving any overhanging shadow of darkness.

Both judge of the world by their own time and circumstances, just as we are too apt to judge of each other by ourselves.

A due regard to the psychonomy of nations would throw much light upon many abstruse points of history, and often serve to corroborate narrations which appear marvellous and incredible to us. Thus, as we have for the most part,* left off eating human flesh in these

* We write thus reservedly because there are some well attested recent instances of cannibalism in Ireland. The following anecdote is likewise narrated by Leyden. "Reiterated complaints having been made to James I. of Scotland, of the cruelties of the Sheriff of Mearns, James exclaimed, 'Sorra' gin the Shirra' were sodden, an' supp'd in broo'.' Thereupon four Lairds decoyed the Sheriff to the top of the hill of Garrock, and having prepared a fire and a boiling cauldron, they plunged the unlucky man into the latter. After he was *sodden* for a sufficient time, the savages fulfilled to the letter the

islands for some thousand years or more, historians reject as utterly incredible that our forefathers were cannibals; and some still more tender-hearted philanthropists even venture to assert that cannibalism has not and never had an existence anywhere. Whereas if they would compare the evidence with the psychonomy of the nations of whom the circumstance is narrated, instead of with our own, they would instantly perceive in it nothing unnatural nor incredible. Thus also infidel writers, unable to comprehend the fervent and assured hope of a blessed immortality which supported the martyrs, deny, as repugnant to human nature, the patient sufferings of the early Christians. And thus again commentators on the Bible, both infidel and credent, have made sad havoc of many texts, by endeavouring to interpret them by European manners and habits. This

King's hasty exclamation by *supping the shirra-broo!*" If the subject were more agreeable to dwell upon, it would be easy to furnish many other well-attested instances of the slacking of the hunger and thirst of revenge by a repast of human flesh.

inattention to national psychonomy is moreover a fertile cause of the mal-administration of colonies, and was the root of nine-tenths of the errors in Indian affairs during the last century.

Seeing then the importance of fully understanding the psychonomy of nations before criticizing their records, we should reject no probable key to that important knowledge; and, if physiognomy would furnish such a key, it should be hailed as an important element in historical criticism. This consideration has induced us to complete our system by a few remarks on National Noses. For no part of the physiognomy is more important to be comprehended than the Nose, if Nasology is correct; because the mental faculties which it pourtrays are more important than those revealed in the other features; and because, being immoveable and permanent in its outline, the artist gives us its national or individual form, without the distortion which the action or passion exhibited may make it necessary to throw over the other more pliant features.

Reserving then till a future chapter any

further observations on National Noses, we will now consider a few individual instances of the Roman Nose.

This Nose is common to all great conquerors and warriors, and other persons who have exhibited vast energy and perseverance in overcoming great obstacles without regard to personal ease, or the welfare of their fellow-men.

The following have pure, or very nearly pure, Roman Noses :—

Rameses II.*
Julius Cæsar.
Henri Quatre.
Charles V. of Spain.
Duke of Wellington.
Canute.
Gonzalo de Cordova (the Great Captain.)
William III.
Sir W. Wallace.

---

* Supposed to be Sesostris, the Shishac of Scripture, at all events a great warrior as appears from Egyptian sculptures, from which his Nose is ascertained.

Robert Bruce.
Queen Elizabeth.
Edward I.
Columbus.
Sir Francis Drake.
Cortez.
Pizarro.
Washington.
Henry VII.
Cato the Censor.
Earl of Chatham.

The well-known, because (as their Noses likewise attest) strongly-marked, characters of these persons makes it unnecessary to allude even briefly to their biographies. Their names are sufficient to bring at once before the mind their energetic, persevering and determined characters. They were persons whom no hardships could deter, no fears daunt, no affections turn aside from any purpose which they had undertaken; that purpose being (from the absence of the Cogitative) always of a physical character; and (from the absence of the Greek) always pursued with a stern and reckless disregard of their own and others' physical ease and welfare. Their successes were attained by energy and perseverance, not by forethought and deep scheming. They were not the men of the closet, but of the field. Physical action, not mental activity, was their adopted road to success. For this reason, and because history is little more than a chronicle of physical action, wars and bloodshed, the owners of Roman Noses occupy the largest portion of their fellow-men's thoughts and of the historical page.

The ancients acknowledged the foregoing Nasal Classification, for they represented Jupiter, Hercules, Minerva bellatrix, and other energetic Deities with Roman Noses, while they gave pure Greek Noses to the more refined Apollo, Bacchus, Juno, Venus, &c. The debased and unintellectual Fawn and Satyr they pourtrayed with Snub or Celestial Noses; thus imparting to their countenances the low cunning or bestial inanity appropriate to those mythological inventions.

It must not, however, be inferred from the majority of warriors' names in the above list, that the Roman Nose necessarily indicates a warrior.

These names are only selected because they afford well-known and easily verifiable instances, requiring neither pictorial nor biographical illustration. Energy may be equally conspicuous in any other department of life, and display itself as fully in the civilian as in the warrior. Two of the individuals adduced are striking instances of this. They were men of remarkable parallelism of character, and, though differing in other facial features, their Noses were very similar. Cato the Censor and the Earl of Chatham.

CATO, THE CENSOR.
(From a gem in the Florentine Museum.)

The events of their early life—those events

which always bear most clearly the impress of the mind, because actuated by choice and not by circumstances, or regard to consequences—were almost identical. They both entered the army in youth, and both quitted it for the Senate. Here each displayed those powers of eloquence which raised them to the highest eminence, and will transmit their names to the latest posterity. Its peculiar feature was that energetic, powerful, and determined vehemence of language, which takes the mind prisoner and carries the judgment with it by storm. It was irresistible. Before it all minds of less power, though of greater intellect and activity, recoiled. The orations of Cato are unhappily lost. But Cicero, a master of eloquence, and well enabled to compare them with similar compositions, passes upon them the highest eulogiums. The eloquence of Cato has been compared for its force and energy to the eloquence of that Demosthenes before whom Philip of Macedon quailed, and whose tremendous orations have given the name of Philippics to all sarcastic and vehement invectives. Of Chatham's eloquence, it has been said by Wilkes; "Nothing could withstand the force of that contagion. The fluent Murray has faltered, and even Fox shrunk back appalled from an adversary 'fraught with fire unquenchable,' if I may borrow the expression of our great Milton. He had not the correctness of language so striking in the great Roman orator; but he had the *verba ardentia*, the bold glowing words."

Cato led victorious armies into the field, and proved himself an able general; for in Rome the functions of the general and the statesman were united in the person of the Consul.

It became not, however, the Secretary of State to lead armies in person; but while Chatham administered the affairs of this country, "victory crowned the British arms wherever they appeared, both on sea and land; and the four years of the second administration of Mr. Pitt, are four of the most glorious years in the history of the eighteenth century."*

In their retirement they were alike; for neither regarded with complacency the pursuits

* Pict. Hist. of England.

of literature; they required some physical activity in their very idleness, and gardening was the favourite occupation of both. Cato displayed his disregard and even hatred for literary refinement by advising the Senate to dismiss the Grecian Ambassador Carneades promptly, lest his eloquence should corrupt the Roman youth with a love for Greek learning and philosophy.

He cultivated his farm and garden with great skill, and wrote a work on the subject, entitled "*De Rustica.*" Chatham was a landscape gardener of no mean pretensions. He assisted Lord Lyttleton in laying out the celebrated park and grounds at Hagley; and Bishop Warburton eulogizes his skill in gardening as inimitable, and far superior to that of the professor Capability Brown. Not even obedience to the King's mandate could draw Chatham from his country retirement at Hayes.

Neither ever thought he had done serving his country while life lasted, even when bodily health and strength were gone. At eighty-four years of age Cato went on an embassy to Carthage; and Chatham, worn out by the gout and wrapped in flannels, never neglected to take his seat in the House and electrify it with his eloquence when any important question affecting the interests of the country or the liberty of the subject arose.

Notwithstanding their many virtues, they were both coarse-minded, violent men; proud, self-willed, and regardless of the common courtesies and even decencies of society. Both were perhaps indebted for some of their fame to the successful practice of the vice which has been happily designated, as the deference paid to virtue.

It is not, therefore, only in the peculiar circumstances of his death that Chatham resembles Cato, with whom he has therein been frequently compared.

It will be remembered that after Cato's return from Carthage, (the inveterate enemy and most powerful rival of Rome,) Cato, then in the eighty-fifth year of his age, and the last year of his life, never spoke in the Senate without expressing his conviction of the dangerous power of Carthage, and concluding with the celebrated words "*Delenda est Carthago.*" Chatham, when peace with America was proposed on terms which he thought dishonourable to his country, expended his last strength in opposing it, and fell, to survive but a few hours, senseless on the floor of the House of Lords.

As by far the majority of persons have compound Noses, and as their consideration will therefore throw additional light upon the system, we shall add a few observations upon some of them.

The Roman Nose may be compounded with Classes II. and III., rarely with IV.; seldom or never with V. and VI.*

Compound $\frac{\text{I}}{\text{II}}$.—The Romano-Greek Nose.†

The following are instances of Noses of this sub-class:—

* The indications of I. being so decidedly opposed to those of V. and VI., it seems almost impossible for them to be associated.

† The class placed first in these compounds is that which predominates.

Alexander the Great.
Constantine.
Wolsey.
Richelieu.
Ximenes.
Lorenzo de Medici.
Frederick II. of Prussia.
Alfred.
Sir W. Raleigh.
Sir P. Sidney.
Napoleon.

Associated with much physical energy (I.), these persons all exhibited much refinement of mind, a love for Arts and Letters, considerable astuteness and capacity of scheming; (II.) they saw far and quickly, though deficient in deep philosophical powers of thought.

A rather more extended notice of some of the members of the sub-classes will be requisite; as, of course, their characters were less developed, and therefore less known, than those of the pure Classes; but principally in order to point out the more minute touches and, apparently, inconsistencies of character which illustrate the compound form of Nose.

CONSTANTINE.

(From a gem in the Florentine Museum.)

CONSTANTINE, having by a felicitous union of enterprise and cunning, procured his elevation to the Imperial throne, and having defeated the last of his rivals to that splendid dignity, directed his attention to the concentration rather than the extension of his enormous empire, and sought, by building Constantinople, to divert the minds of the people from foreign war and intestine discord; while he at the same time fostered and encouraged the arts by the magnificent decoration of the new capital, to which he brought from Asia and Greece some of their most splendid productions.

Vigorous in war and active in peace, Constantine united all the characteristics of the Roman and the Greek. In war he successfully opposed both civil and foreign enemies, and made himself master of the most extended empire Rome had ever designated by her name. While in the vigour of his age, he moved with slow dignity, or with active vigilance, according to the various exigencies of peace and war, along the frontiers of his extensive dominions, and was always prepared to take the field either against a foreign or a domestic enemy.

But when he had gradually reached the summit of prosperity and the decline of life, he became sensible of the ambition of founding a city which might perpetuate the glory of his name, and he then exhibited all the capacities for the enjoyment of the luxuries of peace which had hitherto lain dormant in his mind. The mere building and fortifying a city, which would have satisfied the ambition of the coarser-minded Roman, was not his ambition only. He desired to decorate it with the highest efforts of human genius, and make it not only a monument of his military prowess, but also of his taste and refinement. For this purpose he founded schools of architecture to supply the disparity which his fine taste detected between the degenerate artists of his time and those of early Greece. The immortal productions of Phidias and Lysippus were dragged from other countries to adorn his capital; and, unmindful of the injustice, he despoiled the cities of Greece and Asia of their most valuable ornaments. The trophies of memorable wars, the objects of religious veneration, the most finished statues of the gods and heroes, of the sages and poets of ancient times contributed to the splendid triumph of Constantinople.*

The character of WOLSEY was very similar to that of Constantine. We might almost venture to assert that had he been placed in the same situation he would have pursued the same course. Yet the only part of their physiogno-

---

* Gibbon.

mies which assimilates are their Noses. One remarkable circumstance in the early life of each identifies the two men and exhibits in them the union of energy with acute tact. Constantine, half assured of his elevation to the Imperial throne, if he could join his father's army and be present with him in case of his death, and having with difficulty obtained permission to visit his father from Galerius, (who dreaded the same event, and delayed the permission, until he believed it would be impossible for him to accomplish his object), travelled post through Bithynia, Dacia, Thracia, Pannonia, Italy and Gaul with such speed that he reached Boulogne in the very moment when his father was preparing to embark for Britain, accompanied him, and finally, by military election, succeeded to his share of the Empire.

When Henry VII. was looking out in his old age for a rich wife, he despatched Wolsey, to whom the vista of future eminence was just opening, to Flanders to treat for the hand of a Princess of the Empire. Wolsey, conscious that in such affairs old age brooks no delay, started on his journey and had returned before the King knew that he was gone. By similar energy and shrewd scheming in pursuit of his own aggrandizement, very analogous to that by which Constantine secured the purple, Wolsey elevated himself to the highest station in his country, and then directed his mind rather to the extension of learning, the encouragement of art, the erection of splendid buildings, and the increase of domestic magnificence, than to an imitation of the warlike pursuits of the ancestors of his monarch; although the disposition of the latter strongly tended in that more physically energetic direction. The noble hall and chapel at Hampton Court and the remains of the colleges which Wolsey founded, still attest his magnificence, his taste, his liberality and his respect for learning.

RICHELIEU was another Wolsey. It is a remarkable fact that the point of identity in actively seeking their own aggrandizement, which has been noticed between Wolsey and Constantine, occurs also in the early life of Richelieu. Having, from interested motives, abandoned the army (for which he was originally destined) for the Church, and the Pope having refused, on account of his extreme youth, to sanction his elevation to the Bishopric for the sake of which he had taken orders, he resolved to overcome this difficulty in person; and setting off for Rome, gave the Pontiff such convincing proofs of his talents that he was consecrated Bishop forthwith at twenty-two years of age, and thus laid the foundation of his future eminence.

He conducted in person the siege of Rochelle, and baffled the finest military geniuses of Europe; he out-intrigued the ablest diplomatists; he nourished arts and commerce, and for the better promotion of learning he founded the French Academy.

In the union of energy of character and refinement of tastes the three celebrated Cardinal-ministers of England, France, and Spain, strongly assimilated.

The anecdotes which have been related of the energetic carving out of their own fortunes by Constantine, Wolsey and Richelieu, find also their parallel in the early career of XIMENES. The son of noble parents, but without wealth or patronage, he had nothing but his talents and the energy of his character to carry him successfully through life. He began as a student at Salamanca; but finding that sphere too limited for his ambition, he undertook a journey to Rome, where he soon distinguished himself as an advocate, but preferring the Church, took holy orders.

Sixtus IV. had bestowed upon him the reversionary grant of the first benefice which should fall vacant in Spain. This proved to be Uceda; and, on the demise of the incumbent, he produced his letters, and took possession with such promptitude and despatch that he baffled the Archbishop of Toledo, who considered the benefice to be in his gift, and had promised it to one of his dependants.

Like Richelieu he took the field in person, and in spite of the jealousy of the King, the dissensions of the generals, and the mutiny of the soldiers, he succeeded in taking the town of Oran on the coast of Barbary; the first success of any moment which the Spanish army could boast in a campaign of four years' duration.

He devoted himself, in after life, to the encouragement of popular education and the advancement of higher learning, in no less degree than his brother Cardinals before named. He founded a school for the education of the daughters of the poorer nobility, and subsequently provided them with marriage portions.

He established the University of Alcala, richly endowed it, and filled its professorial chairs with the most distinguished learned men of Europe. Here he undertook the magnificent work, known as the Complutensian Bible. It was the first Polyglott Bible ever published, and as such affords a striking contrast to the otherwise undeviating opposition which Spain has offered to the spread of true Christianity and the circulation of the Scriptures.

It should, however, be remembered that even this was a sealed book to the laity, since it did not comprise a version in the vernacular. It contained the Old Testament in the Hebrew, the Septuaguint, the Vulgate of St. Jerome, and the Chaldee Paraphrase with Latin translations, and the New Testament in the Greek and Vulgate.

It was the work of fifteen years, and when the last volume was brought to Ximenes, shortly before his death, he exclaimed: "Many high and difficult matters have I carried on for the State, yet is there nothing which I have done, that deserves higher congratulations than this edition of the Scriptures; the fountain-head of our holy religion, whence may flow purer streams of theology than those which have been turned off from it." The whole cost of the work, fifty thousand gold crowns, was defrayed by Ximenes.

In LORENZO DI MEDICI, we meet with another of those characters, frequent among men eminent in public affairs, which unite refinement of taste with physical energy. To live in the world's eye with success, it is necessary to exhibit something *ad captandum vulgus.* There must either be the intense energy of the Roman, or the more moderate energy with the taste and magnificence of the Romano-Greek. Hence, while the former class of Nose prevails among those who have won fame and honours by arms merely, the latter is frequent among those who are chiefly celebrated for their statesmanship. But both energy and statesmanship were necessary to him who would secure a world's fame as ruler of a petty Italian State. The head of a State too weak to be feared in war, and too turbulent to be governed in calm tranquillity, required some other qualities besides energy, in order to be respected and honoured by his cotemporaries. These qualities were happily united in Lorenzo di Medici. Firm in danger, prompt in action, lavish in expenditure, refined in taste, accomplished in learning, expert in art, he was every way formed to win laurels in an age which boasted the greatest statesmen, the best artists, and the most profound scholars. The vigour and promptitude with which he repelled the celebrated conspiracy of the Pazzi family, hanged an Archbishop on the spot in full canonicals, and punished the conspirators, alone attests his energy. The title of Magnificent which he earned in an age celebrated for its magnificence, demonstrates his lavish liberality; while his love for antiquities, his patronage of the arts of sculpture and painting, his studious devotion to learning and the writings of the ancients, bespeak the refinement of his mind. Among other institutions, he founded a school for the study of antiquities and furnished it with the finest specimens of ancient workmanship. "To this institution, more than to any other circumstance, we may, without any hesitation, ascribe the sudden and astonishing proficiency, which, towards the close of the fifteenth century, was evidently made in the arts, and which, commencing at Florence, extended itself to the rest of Europe.

"'It is highly deserving of notice,' says Vasari, 'that all those who studied in the gardens of the Medici, and were favoured by Lorenzo, became most excellent artists, which can only be attributed to the exquisite judgment of this great patron of their studies.'"*

FREDERICK II. is another example of the union of refined tastes with vigorous energy. It is not so much for his military genius that he is to be remembered and respected, as for the impulse he gave to Prussian intellect, and thence generally to German mind.

* Roscoe's Life of L. di Medici. Chap. IX.

It is true this was hardly perceptible till the present century, for until the peace of 1815, Germany had been the seat of almost incessant warfare, and was, therefore, disabled from pursuing the arts of peace with success. But thirty years' peace has enabled her to perform great things, and to justify a pretty sure hope of yet greater. We ought to be far in advance of her, for where she now is we were exactly two hundred and fifty years and upwards ago. Till the reign of Elizabeth, England had been, like Germany till 1815, the seat of perpetual war or religious discord. At the end of the sixteenth century in England, and at the beginning of the nineteenth in Germany, the Teutonic mind began to develop itself with effect. The same deep investigations in history, the same subtle disquisitions in metaphysics, the same love of philological criticism that distinguished English literature in the early part of the seventeenth century belong to German literature in the nineteenth, and are combined with the same coarseness of manners that marked our ancestors. The Germans, still delight in those rude, indecent productions, called Miracle-plays or Mysteries,* which amused the predecessors of Shakspere: legalized wager of battle, semi-feudalism, masks of fools dancing in a gigantic beer-barrel and chanting the praises of beer, deer-battues, perpetual duelling and beer-swigging, millions pilgrimaging to the Coat of Trèves, the implicit reception of sham Miracles, all mark a state of society little removed from that magnificent barbarism which stained the rush-strewn court of the ear-boxing and swearing Elizabeth.

In refinement, and that wealth which springs from Science, we have advanced far beyond Germany; but in that wealth which emanates from Mind we are only on a par with her. The causes of this will be considered more fully hereafter, when we treat under Class III. of the causes of the decline of Wisdom.

The impulse given to German mind may in

* See Hone's description of one performed in 1815 before several crowned heads of Europe for three successive days; *Hone on the Mysteries.* See also *Wilhelm Meister*, Vol. I.

a great measure be attributed to the pains which Frederick II. took to civilize and educate his people. For this purpose he founded numerous popular schools, it is said as many as sixty in one year. He instituted an Academy of Sciences and fostered Universities. He patronized Commerce and the Arts, and by his wise administration as much as by his military talents raised Prussia to the rank of a second-rate European State. The military success of the correspondent of Voltaire, it is unnecessary to do more than refer to.

Machiavellism formed a strikingly distinctive feature in the characters of all the foregoing personages. They all possessed more of the wisdom of the serpent, than of the innocence of the dove. It may be thought, however, that we employ too strong a term in calling this Machiavellism. A less strict morality would only call it policy, worldly wisdom. In men of strong conscientiousness, astuteness may be little or nothing more; but where the moral sense is weak, it easily passes into duplicity and dishonest craft.

The shrewd policy and worldly wisdom by

which the great ALFRED civilized a barbarous people, and tamed to quietude a nation of turbulent robbers, has never been accused of departing from a strict morality. It may be that he is somewhat indebted to the partiality of the monkish historians for the very flattering pictures of him handed down to us. The prompt and energetic manner in which, from time to time, he fell upon and defeated the Danes who ravaged the country is too well known to need mention, and the prudent means by which he endeavoured to incite his people to educate themselves has been often the subject of praise. In a remarkably illiterate age, he alone courted literature, and, conscious of its power to civilize his people, urged them to follow his example. Nevertheless, he did not forget the more arduous duties of a King. While devoting a large part of his time to learning, he never neglected the interests of his country; nor suffered her liberties to be trampled upon by invaders while he was cultivating the arts of peace. His biographer, quaintly and somewhat poetically, describes the King's studious mind and gubernatorial talents. "Like a most productive bee, he flew here and there asking questions as he went, until he had eagerly and unceasingly collected many various flowers of Divine Scriptures, with which he thickly stored the cells of his mind. His friends would voluntarily sustain little or no toil, though it was for the common necessity of the kingdom ; but he alone, sustained by the divine aid, like a skilful pilot, strove to steer his ship laden with much wealth, into the safe and much-desired harbour of his country though almost all his crew were tired, and suffered them not to faint or hesitate, though sailing among the manifold waves and eddies of this present life."*

The circumstances in which men are involuntarily placed marvellously affect their actions. Crowd together a number of young trees in one small plot, and how slowly they grow, how stunted they become! Remove them to separate stations, where their roots may spread, their branches expand, and their leaves drink freely of the sun and air, and how soon

* Asser's Life of Alfred.

they take their place among the giants of the forest. So it is with men. Crowded in cities, undistinguished by birth, and unassisted by patronage, many a hero dies unseen and unnoticed—

> "Some village Hampden, that with dauntless breast,
> The little tyrant of his fields withstood;
> Some mute inglorious Milton here may rest,
> Some Cromwell guiltless of his country's blood."

Let it not, therefore, be imagined, from the foregoing instances, that every Greco-Roman Nose indicates an energetic statesman, or a literary monarch; or that the same actions are to be predicated from the same form of Nose in different men under different circumstances.

Energy and refinement may exist in every department of life. The peasant may furnish as illustrious an example of either as the Prince. But what a King has, these heroes want; and so they die unhonoured for lack of a record. The illustrations are, therefore, necessarily drawn from the high and mighty of various spheres.

Stars of lesser magnitude, however, present themselves to shed a further light upon the subject.

SIR WALTER RALEIGH and SIR PHILIP SIDNEY were two men whose characters exhibited many points of identity.

In any arduous enterprize which promised fame and honour, Sir Walter Raleigh was always prominent. Eager to support the Reformation, he served in the Protestant army as a volunteer during the civil wars in France, and afterwards tendered his services to the Netherlands in their contest with Spain for civil and religious liberty. One of the most attractive enterprizes of the reign of Elizabeth to men of energy and forethought was, however, that presented by the recently-opened field of American discovery. Into this Raleigh threw himself heart and soul. With his half-brother, Sir Humphrey Gilbert, he made the then perilous voyage to the New World, but failed to establish a firm footing on its shores.

Still he was not to be thus foiled. After a careful consideration of the best authorities, he came to the just conclusion that there was land north of the Gulf of Florida, a tract then wholly unexplored. Having obtained from the Queen the inexpensive grant of all he might discover, be it sea or be it land, be it inhabited or be it void, he fitted out vessels of discovery; and, though not permitted by the wary Queen to accompany them himself, they verified his predictions by discovering the country now called Virginia—a name which the virgin Queen herself bestowed upon it.

But it was not by his energy that Raleigh alone distinguished himself. The young Protestant volunteer, and the American adventurer would long since have been forgotten among a host of compeers, had not he presented far higher claims to the notice of posterity. "Raleigh was one of those rare men who seem qualified to excel in all pursuits alike; and his talents were set off by an extraordinary *laboriousness*, and *capacity* of application. ($\frac{1}{11}$). As a navigator, soldier, statesman, and historian, his name is intimately and honourably linked with one of the most brilliant periods of British history."*

Sir Walter Raleigh occupies a distinguished place in literature, both as a poet and an historian. It is probable that only a small portion of his poetry has come down to us. He seems to have regarded it but lightly himself, and many very beautiful pieces, which there is no reason to doubt owe their origin to his creative brain, are without name, and only preserved in some obscure miscellaneous collections, under the modest signature 'Ignoto.' One of these, sometimes entitled "The Lie," and sometimes "The Soul's Errand," is as beautiful, as Christian, and as philosophic a poem as any in the language; yet so little pains did he take to secure to himself the literary fame of the words with which he had relieved his labouring soul, that it has been attributed to divers poetasters, and, among others, to that most wretched inharmonious scribe, Joshua Sylvester.

* Life of Raleigh, 6 Port. Gal. p. 10.

Spenser eulogizes Raleigh's poetic powers as those of one

> ". . . . as skilful in that art as any."*

He likewise entitles him 'the summer's nightingale,' and hints that he had in store a poem on Queen Elizabeth, which might rival "The Faerie Queene:"—

> "To taste the streames, that like a golden showre,
> Flow from thy fruitful head, of thy Love's praise—
> Fitter perhaps to thunder martial stowre—
> When so thee list thy lofty Muse to raise;
> *Yet till that thou thy poem wilt make known,*
> Let thy faire Cynthia's praises be thus rudely shown."

But poetic effusions are not the only contributions of Raleigh to literature. During his long confinement in the Tower, on charge of treason, he relieved his solitude by compiling a "History of the World;" an undertaking sufficient to appal the most active and learned man under the most favourable circumstances, but which appears something superhuman when

* Colin Clout.

attempted and almost accomplished by a wretched prisoner lying under an unjust sentence of death.

This History commences at the Creation, and descends as far as the end of the second Macedonian War; when, in consequence of the death of Prince Henry, for whose instruction it was intended, he ceased from his arduous labours. The work displays a vast extent of reading in history, philosophy, theology and Rabbinical learning.

Like Raleigh, Sir Philip Sidney combined the characters of the warrior and the author. His Arcadia was a work of poetic prose, better suited to the time in which he lived than to any subsequent period, and is almost forgotten; and the stiffness and hard formality of his poetry has almost sunk it in like oblivion. A writer who is not an author for all time, may be a very useful and agreeable one in his day, but lacks power and thoughtfulness. It is only those who have the "one touch of Nature which makes the whole world kin," that are independent of time, and live with the kindred spirits of all ages.

Time puts out the lesser lights which burn only to light some small apartment and corner of the world, but cannot extinguish the suns which are formed to illuminate the whole earth.

Sir Philip Sidney was rather a discerning patron of letters than a man of letters. He was the first patron and friend of Spenser, whom he introduced to the Queen, and their friendship endured till Sidney's lamented death. Perhaps in the whole range of literary history, there is no incident so beautiful as the mutual friendship and familiar intercourse of Raleigh, Spenser and Sidney. This pleasing friendship is frequently alluded to by Spenser. The 'Faerie Queene' is dedicated to Raleigh, whose return from his Western Expedition is celebrated in the Pastoral entitled, "Colin Clout's come home again;" from which we learn that it was their custom to recline

> ". . . . amongst the coolly shade
> Of the green alders by the Mulla's shore."

and recite to each other their poetic effusions.

How beautiful a picture of the simplicity of

great minds! It strikes us as a more lovely picture than the much-admired one of Chaucer, solitary among the daisies of the Woodstock meadows.

Sidney inspired Spenser with no mere mercenary friendship, the affection of the client for his patron's substantial marks of favour. When death smote Sidney on the sad field of Zutphen, Spenser invoked every Muse to weep over his untimely fall, and celebrated his virtues in the beautiful elegy "The Tears of the Muses for Astrophel." It will perhaps relieve the dryness of our subject, to observe that the first poetical use of the Forget-me-not, (*Myosotis palustris*) as a symbol of faithfulness, occurs in this poem, and the English reader may there find a more fitting reason to esteem this little flower than the absurd German legend of a drowning knight throwing a spray of it to his ladye-love.

The Astrophel of the following lines from Spenser's Elegy, is Sidney; Stella is the name by which Sidney addressed his Mistress, who, it is feigned, was unable to survive his loss, and,

". . . . followed her mate, like turtle chaste,
To prove that death their hearts cannòt divide,
Which, living, were in love so firmly tied.

"The Gods which all things see, this same beheld;
And pittying this paire of lovers trew,
Transformed them, there lying on the field,
Into one flowre that is both red and blew.
It first growes red, and then to blew doth fade,
Like Astrophel, which thereinto was made.

"And in the midst thereof a starre appeares,
As fairly formed as any starre in skyes,
Resembling Stella in her freshest yeeres,
Forth darting beames of beautie from her eyes;
And all the day it standeth full of deow,
Which is the teares, that from her eyes did flow.

"That hearb of some, Starlight is call'd by name,
Of others, Penthia, though not so well;
But thou, whenever thou dost find the same,
From this day forth doe call it Astrophel.
And whensoever thou it up doost take,
Doe pluck it softly for that shepheard's sake."

May the injunction of the last lines never be forgotten by any one who knows that the forget-me-not is associated with the friendship of two such noble-minded men!

It is hardly necessary to say that Sir Philip

Sidney fell gallantly fighting at the battle of Zutphen, or to narrate the interesting anecdote of his refusing a drink of cold water till a wounded soldier had partaken of it, saying, "Thy necessity is yet greater than mine;" thus nobly displaying both firm endurance and sensitive humanity.

The other instances, ALEXANDER THE GREAT and NAPOLEON, may be best treated of by contrasting them with their opposites; and we shall thus be enabled to illustrate, at the same time, both the Roman and the Greek Noses more fully. Moreover, while the contrast will clearly demonstrate the distinctive characteristics of those Noses, it will also evince how important it is to attend to compound forms, and how materially the character is affected by the intermixture of classes.

Of all the conquerors whose wild ambition has stained with blood the page of History, Alexander and Napoleon alone fought from a high romantic motive—the desire of eternal fame. By virtue of a large share of the Roman Nose, they pursued their favourite and chosen career with determined energy and a reckless disregard for the lives of others; nevertheless, being strongly gifted with the Greek, they might in some other sphere have been high artists of some class; but having the sword in their hands they pursued intellectual fame by its means.

It is difficult to say whether the Roman or the Greek form predominates in their noses; for they are perhaps as much Greco-Roman as Romano-Greek; but as they were warriors, we place them here because it will be advantageous to draw an illustrative contrast between their characters and noses, and the characters and noses of too many other mere conquerors, whose noses have been purely Roman.

Let us briefly contrast Julius Cæsar and Alexander. They were both, in the prime of life, placed at the head of a large empire, firmly seated, with a large army and all the world open to their grasp. Their Noses alone differed. Alexander while pursuing everlasting fame by his arms, and earning what was then deemed the highest glory, steadily devoted himself to the extension of scientific knowledge. Under his revered master Aristotle, he acquired much learning, and, when he ascended his father's throne, devoted his arms as much to the conquest of the then unknown realms of science as of the kingdoms of the earth. His army was always accompanied by learned men, whose sole

JULIUS CÆSAR. ALEXANDER THE GREAT.

(From gems in the Florentine Museum.)

duty it was to investigate the history, religion, and arts of the countries he passed through, to collect rare animals and plants, statues, coins, and objects of art or curiosity to be transmitted to Greece for the study of his master Aristotle. It has been well said, "If there had been no Alexander, there would have been no Aristotle." We do not laud the man who sought glory by the destruction of others, but merely assert that, as these acts prove, his motive to arms was a high intellectual one, and consistent with the compound character of his Nose.

Look at Julius Cæsar on the other hand. Under similar circumstances, what was his ambition? To make himself imperial master of Rome, and to subject his fellow-citizens for his own personal aggrandizement. His thoughts never extended beyond his own petty existence. Posterity never entered into his calculations. Unlike his successor Augustus—though he had greater facilities if he had been less sensually ambitious—he patronized no art—literary or scientific. His one idea was self, without one refinement or softening alloy. Granted that Alexander's ambition was also selfish, there was yet this difference between them; the one (Cæsar) sought only his *present personal* and *sensuous* profit; the other (Alexander) laboured to earn "a name on History's page to make him 'GREAT.'" The one was the common prose, the other the epic poem. The one sacrificed his fame to himself, the other himself to his fame; and the world has recognized and recorded this distinction; for while the one is remembered as "the enslaver of his country," the other is immortalized as "the Great."

A similar contrast may be drawn between the characters and noses of the two modern heroes, Napoleon and Wellington. Like Alexander and Cæsar, the only point in which their characters assimilate is their warrior, physical energy; and this exhibits itself in whatever is Roman in their Noses. In all other respects they are diametrically opposite; the Nose of Wellington being purely (almost in excess) Roman; while Napoleon's partakes largely of the refining qualities of the Greek.

To describe the character of Napoleon would be to repeat what we have said of Alexander; for whether the similarity was accidental, or arose from mental conformity (their Noses were remarkably alike), or was intentionally imitative on the part of the former,* it is certainly most striking.

Ambition of future fame was far more the ruling passion of Napoleon than lust of present power. His mind, with all its imperfections and meannesses (as whose is without?), was too noble to be satisfied with mere personal aggrandizement.

All the great mistakes of his life were occasioned by his obedience to the passion for future fame. When swayed by the mere desire of power, all his acts were successful. As General, Consul, or Dictator, Napoleon never made one false step; but when he became Emperor, when he saw all Europe (except one little pugnacious island) lying helpless at his feet, he began to revolve schemes which could not enhance, but might risk, his personal power. Then he attempted to realize his long-cherished dream of Eastern conquest—a conquest not to be held, but to be overrun; a conquest like that

* If Napoleon was an imitator of Alexander, it was only another point of identity between them; for Alexander was an imitator of Bacchus.

of Alexander, Nadir Shah, or Kinghis Khan. Often and often did he exclaim, " the seat of all fame is the east." To realize this empty fame, he took the false step of invading Egypt. Foiled there, he still hoped to penetrate Asia by land, and gathered all his strength to overwhelm Russia, his last and greatest error. They greatly err who think these were mere schemes to keep France embroiled, lest peace should annihilate his power. They equally err who ridicule and attribute to a childish vanity his ambition to link himself by marriage with the imperial families of Europe. It was no childish vanity, but a politic endeavour to found a dynasty, which should hand down his name as its founder to the latest ages. They again who can see nothing better in the melancholy spectacle of Napoleon at St. Helena, engaged in falsifying records and altering figures to deceive the world, but a drivelling vanity, utterly mis-comprehend the man. Fame, fame to the utmost limits of human duration was to his last moment his highest ambition. Foiled in every thing else, he yet hoped to secure fame. He knew that under his name the most eventful page in the History of Europe, since the fall of Rome, must be written, and he naturally desired

> "To be among the worthies of renown,
> And so sit fair with fame, with glory bright."
>
> DANIEL.

To describe the character of Wellington, is to reverse that of Napoleon. Napoleon was shrewd, artful, and deceitful; Wellington open-hearted, strong-sensed, candid and sincere. Napoleon a clever statesman; Wellington obtuse in politics. Napoleon a great strategist; Wellington short-sighted, though daring, in the field. Napoleon a lover and patron of arts; Wellington a despiser of them. Napoleon said to be personally timid; Wellington constitutionally brave. Napoleon's cruelties were acts of cool calculation and state-policy; Wellington's of military fury. Napoleon poisoned his prisoners because he did not know what else to do with them, and murdered the Duke d'Enghein to produce "an effect" in Europe; Wellington's cruelties were the necessary consequences of war energetically carried on, and were never the result of cold-blooded predetermination.

Before closing this section, we would request the reader's attention to the strong proof of the truth of the hypothesis derivable from the fact that like Noses, with like circumstances, (*cæteris paribus*, as the phrenologists say) produce like characters: for instance, Wolsey, Richelieu, Ximenes, Lorenzo di Medici, Alfred:—Sidney, Raleigh:—Alexander, Napoleon.

CHAPTER III.

OF THE GREEK NOSE.

CLASS II.—THE GREEK, or straight Nose, is perfectl straight; any deviation from a right line must b strictly noticed. If the deviation tend to convexity it approaches the Roman, and the character is im proved by an accession of energy; on the other hand when the deviation is towards concavity, it partakes o the Celestial, and the character is weakened. It should be fine and well chiselled, but not sharp.

It indicates Refinement of character; love for the Fin Arts, and *Belles Lettres*; Astuteness, craft, and a pre ference for indirect rather than direct action. Its owner is not without some energy in pursuit of tha which is agreeable to his tastes; but, unlike th owner of the Roman Nose, he cannot exert himself in *opposition* to his tastes. When associated with the Roman Nose, and distended slightly at the end by the Cogitative, it indicates the most useful and intellectual of characters, and is the highest and most beautiful form which the organ can assume.

THIS Nose. like the Roman. takes its name

from the people of whom it was most characteristic—physically and mentally. On these two parallel facts (with others of a like kind) much stress may be justly laid, although they are old and trite. But this very triteness is the proof of their truth. It proves that the hypothesis which attributes certain mental characteristics, well known to belong to the Romans to the Roman Nose,—and so of the Greeks to the Greek Nose, and of the Jews to the Jewish Nose,—is founded in nature; and, so far from being a fanciful invention, is a fact long-recognized, and as old as the creation of the human proboscis.

Requesting the reader to bear in mind the form of the Greek Nose and its indications, we would remark how exactly the latter correspond with the character of the ancient Greeks as a nation. It is unnecessary to expatiate on their high excellence in art, their lofty philosophy, their acute reasoning, or their poetical inspiration —these are known to every school-boy. Their craftiness, their political falsehood, and shrewd deceitfulness were celebrated in ancient days as now, and "*Græcia mendax*," "*Danaûm insidiæ,*" were epithets as true and as commonly applied in the time of Augustus, as at the present hour by modern travellers. "*Timeo Danaos et dona ferentes!*" exclaims the cautious Priest of Troy, referring to the well-known character of the treacherous enemy. And what a contrast to anything recorded in Roman warfare does the Trojan War itself exhibit! The Romans would have battered down the walls with their furious engines; the wily Greeks invent a stratagem by which the enemy pull down their own walls. If we may credit Homer—and, if not for the facts, we may for his fine portraitures of Grecian character—there was a vast deal more talking than fighting during the ten years' siege. There was plenty of the *morale,* but very little of the *physique,* as a Frenchman would say. In truth, the contrast between the Romans and the Greeks was as great in the latter as in the former.

The Greeks were no nation of hardy warriors, though they were always quarrelling among themselves in petty battles which have won an undeserved celebrity by the talents of their historians. Were it not for the writings of Thucydides, the Peleponnesian War would rank no higher than the border skirmishes of the Scots and Northumbrians, or the expeditions of the Sioux and Pawnees. A simple geographical fact is sufficient to prove this against all the moral power of the most glowing and eloquent historian. Greece is about one-fourth less than Scotland, and its recorded population was about the same. Is it possible that, in such a corner, a war of three-and-twenty years' duration could be more than a series of skirmishes and predatory expeditions? More than that, must, in a much briefer space, have annihilated the whole population. More than that, and at the end of twenty-three years the States of Greece must have been in the condition of the celebrated Kilkenny cats, which fought till only the tip of the tail of one of them was left. The battles of Marathon, Thermopylæ, &c., against foreign foes rank higher, because they were fought and won under a high intellectual inspiration, entirely consistent with Class II.—the love of country. But with these battles the war ended; the

Greeks did not, as the Romans would have done, follow up the defeat of the enemy with a counter-incursion into his country and an attempt at foreign conquest. He was driven from their territory; their hearths were secure; their gods replaced on their pedestals; their temples re-purified, and that satisfied *their* ambition. The Greeks made no foreign conquests; boasted no extended empire. The wars of Alexander seem the only exception; but of that Monarch himself we have already treated, and of his battles it may be said, that they were not fought by Peleponnesian Greeks (of whom we are now speaking) but by Macedonians and Asiatic mercenaries; who were in all probability—though it would demand a volume on ethnography to prove it —a wholly different race.

But, if we were only prepared to substantiate our hypothesis by these general facts of national characteristics, it would be very unsatisfactory; as it is obvious that nothing could be easier than to manufacture and support a theory by moulding it to a single general fact. It is by the multiplicity of isolated *indi-*

vidual cases that the hypothesis must stand or fall. And we are happily in a position again to adduce these in its favour.

The following persons will, on an examination of their portraits, be found to have possessed Greek noses :—

Petrarch.
Milton, (in youth.)
Spenser.
Byron.
Shelley.
Boccacio.
Canova.
Raffaelle.
Claude.
Rubens.
Murillo.
Titian.
Addison.
Voltaire.

It will be perceived that this list (which, like all the others, might be very much extended) contains the names of poets and artists of the highest *beauty* and elegance, though not of the most intense and deepest *thought*.

RAFFAELLE.

Beauty is their highest excellence, their chief praise. Exquisite melody, ætherial fancies, felicitous expression, a fine perception of the Beautiful, as distinguished from the Sublime, whether on paper or canvass, (for it is only the difference in the *mécanique*, or vehicle of expression, which constitutes the difference between the Artist and the Poet), are their best attributes. Addison and Voltaire are the only two of the above instances who never excelled in Poetry or Art, though both assiduously courted the former Muse. Nevertheless Addison is an illustrious instance in our behalf. Is not the *beauty*, the correctness, the euphony of his style still an object of emu-

ADDISON.

lation? Has it not for above a century been the model of good writing? And yet it is too true that nothing equally permanent can be found, which is at the same time so weak and tame in thought, so shallow in reasoning, or so lax in argument. In fact, it owes all its permanency to its euphony, its musical harmony and exactness of expression.

The absence of a noticeable development of the Cogitative (Class III.) accounts for the deficiency of higher qualities in these disciples of the Beautiful. For this reason the Greek nose is more interesting in its compound form, Sub-class $\frac{II}{III}$. the " *Greco-Cogitative,*" than in its simple form.

Of the above instances, Voltaire is the most decidedly deficient in the Cogitative, which is always essential to indicate a capacity for the deep, close and serious thought requisite to constitute a truly great and philosophic mind. The angle at which his nose stood from his face was quite 45°, and therefore much too great to exhibit faithfully the higher characteristics of the Greek. It was, moreover, exceedingly deficient in the broadening property of Class III; and we presume that no one will assert that Voltaire possessed " a truly great and philosophic mind." Surely no man, who ever wrote so much, and on such varied subjects, ever devoted less time to close intense thought. He did not even stop to examine his facts; but, having a brilliant wit and " the pen of a ready writer," he rapidly evolved some fanciful theory, or started some fallacious argument from such unauthenticated data as he happened to be possessed of. All this was indicated by his sharp Greek Nose; for it was acuteness, not depth; readiness, not thought; careless unprincipled wit, not study; attractive style, not sound matter, which earned him his short-lived fame. Hence, Voltaire, though striving all his life to gain the title of philosopher, never succeeded even in the most unphilosophic age and country since the revival of learning, and is now, we believe, wholly excluded from the dignity. It has been truly and wittily said of Voltaire, that "he *half* knew everything, 'from the cedar tree that is in Lebanon, even unto the hyssop that springeth out of the wall,' and he wrote of them all, and laughed at them all."

It will be noticed that the foregoing list contains the name of "Milton, *in youth.*" It is inserted thus, because his portrait, taken *ætat* XXIII, shows that his Nose was not then developed into the Cogitative form which it assumed in later years, when troublous times and anxious cares caused him to reflect profoundly on events around him. Then it expanded at the base and became, like the Noses of all the great men of those stirring times, largely compounded with the Cogitative; under the compounds of which class it will again, at a later period of his life, appear. From this corresponding change in feature with change in character, we might, if we thought proper, demand the same proof for our system which the phrenologists demand for theirs, from the gradual alteration in the skull of the boy Bidder; and though (as our system is, we conceive, better based than theirs) it is unnecessary to lay as much stress upon a single fact as they are compelled to do, yet we think it right not to let this proof pass wholly without observation.

Having already treated at some length of the Romano-Greek Nose (Sub-class $\frac{\text{I}}{\text{II}}$.), it is unnecessary to enlarge here upon its close ally the Greco-Roman $\frac{\text{II}}{\text{I}}$. Of course they are somewhat similar in appearance and character; only as in every compound form, one simple one will generally prevail—Nature, like a bad cook, not always mixing her ingredients in due proportions—it is necessary to distinguish them into different sub-classes.

A noticeable predominance of one form will at once indicate to which sub-class a Nose belongs, and the character will be found to be affected accordingly. Thus a Romano-Greek Nose indicates a more energetic and less refined character than a Greco-Roman. But these are the minutiæ of the science, with which it is not advisable at present to embarrass the reader.

BYRON.

CHAPTER IV.

OF THE COGITATIVE NOSE.

CLASS III.—THE COGITATIVE, or Wide-nostrilled Nose, is, as its secondary name imports, wide at the end, thick and broad, not clubbed, but *gradually* widening from below the bridge. The other Noses are seen in profile, but this in full face.

It indicates a Cogitative mind, having strong powers of Thought, and given to close and serious Meditation. Its indications are of course much dependent on the form of the Nose in profile, which decides the turn the Cogitative power will take. Of course it never occurs alone, and is usually associated with Classes I. and II. rarely with IV., still more seldom with V. and VI. The entire absence of it produces the "sharp" Nose, which is not classified, as sharpness is only a negative quality, being defect of breadth, and therefore indicates defect of Cogitative power.

IT is manifest that without some portion of the Cogitative power, *i. e.*, the capacity of concentrating the thoughts earnestly and powerfully on one focus, no character can be truly great. It is therefore a quality essential to high and durable eminence in every department of life. It matters not what a man's natural talents may be, they will be utterly useless, or worse than useless, if he has not schooled his mind into habits of concentrated thought. It is the want of this severe training which causes so many men of fine talents to be a burden to themselves and others. How frequently have we to lament the humiliating spectacle of a great genius—as the phrase is—flitting about from pursuit to pursuit, without any settled end or aim; now attempting this thing, now dabbling in that; doing all things tolerably well, but nothing perfectly; aiming at everything, but holding fast to nothing; and merely from want of steady settled habits of thought! How melancholy is it to reflect that the want of self-training in early life has converted the blessing of talents into a curse, and turned the fine wheat of Heaven's planting into the rank tares of Hell!

It is from beholding this too frequent spectacle that dull-pated Ignorance repeats with self-complacency the trite proverb, "Geniuses rarely do any good for themselves," professes to despise the talents in which he is consciously deficient, and thanks God that He has not made him a genius.

Begone, thou muddle-pated imbecile! and learn that it is not his genius which has made him what he is, but the want of that in which you equally fail—self-training. Instead of idly despising the noblest gift of Heaven, strive, from his example, to avoid the rock on which he has split, and endeavour by stern, close, severe mental discipline to elevate yourself to a fractional part of the high estate from which he has fallen. Pull him not down to your debasement, but soar upward towards the eminence which he has voluntarily (alas!) abandoned; well assured that though *you* may never reach it, your labour will not have been in vain, and that you may yet place yourself far above the level of the common despisers of genius.

But to our subject—the Cogitative Nose. This Nose long puzzled us. We found it among men of all pursuits, from the warrior to the peaceful theologian. Noticing it more particularly among the latter, we were at one time inclined to call it the religious Nose; but further observation convincing us that that term was too limited, we were compelled to abandon it. We were next, from perceiving it frequent among scientific men, disposed to call it the philosophic Nose; but this was found to be too confined also, as, in the modern acceptation of the term, it seemed to exclude the theologians, and we moreover traced it accompanying other and very different conditions of mind. It soon became manifest, however, that it was noticeable only among very first-rate men (men of the very *highest* excellence in their several departments), and that search must be made for some common property of mind which, however directed by other causes, would always lead to eminence. It appeared to us that this property was deep, close Meditation, intense concentrated Thought, eminently "cogitative" in fact; and, therefore, we adopted this term, which permits to have included in it all serious thinkers, whatever the subject of their cogitations.

It would be wrong to regard it as a mere

coincidence, that, after having from deduction *a posteriori* learnt that this common property is exhibited in the *breadth* of the Nose, we find that if we were, *a priori*, to consider in which part of the Nose a *common* property was to be looked for, we must decide it to be in the *breadth*, for the profile is already in every part mapped out and appropriated to *special* properties.

May we not hail this as one of the beautiful harmonious truths which spring up from time to time, the deeper the subject is investigated, to attest the accuracy of the system? for where by a careful deduction, *a posteriori*, we discover the common property is, there, *a priori*, we perceive it must be in order to act in concert with the *special* properties exhibited by the profile.

To entitle a Nose to rank among the Cogitatives, it should be above the medium between the very full broad Nose and the very sharp thin Nose. The observation is to be confined to the parts *below* the bridge; what may be the properties of breadth *above* the bridge we have not at present observed satisfactorily. It may be remarked as a general rule, that the further a Nose recedes from sharpness the better.

We have said that minds of every bias are found accompanying Cogitative Noses, and this necessarily; for the tendency of the cogitations will be determined by the profile. Thus the Cogitative acts in concert with the other Noses, making useful those qualities which would, otherwise, for ever slumber unknown. The very best Nose in profile may be utterly worthless from defect of breadth; for, as before observed, no talent is of any use without Cogitative power; and every Nose, having breadth as well as length (profile), must be submitted to the test of this Class before a judgment is pronounced upon it. Being, however, anxious to simplify the subject, we have not, in our notices of Classes I. and II., remarked specially on the Cogitative part of their formation, and have reserved until this chapter the instances of those Classes partaking largely of Class III.

In the present brief sketch of the science, however, we shall not attempt to distinguish our instances under the heads of distinct profiles, as, Romano-Cogitative, Greco-Cogitative, &c.; but class together all the compounds partaking sufficiently of the Cogitative form to entitle them to a place among Cogitative Noses.

The following persons have Noses which largely partake of this important formation:—

THEOLOGIANS.	SCIENTIFIC MEN.	LAWYERS.
Wicliff.	Hunter.	Erskine.
Luther.	Jenner.	Blackstone.
Cranmer.	Galileo.	Mansfield.
Knox.	Dollond.	Hale.
Tyndale.	Caxton.	Coke.
Fuller.	Bacon.	Somers.
Hall, Bishop	Whiston.	
Tillotson.	Delambre.	ARTISTS.
Baxter.	Wollaston.	Angelo, Michael
Bunyan.	Smeaton.	Hogarth.
Hooker.	Newton.	
Taylor, Jeremy	Halley.	
South.	Banks, Sir Joseph	
Warburton.	Watt.	

THEOLOGIANS.	SCIENTIFIC MEN.	
Stillingfleet.	Cartwright.	
Chalmers.	Cuvier.	
Priestley.		
Wesley.		

POETS.	STATESMEN AND METAPHYSICIANS.	HISTORIANS.
Homer.	Cromwell, O.	Selden.
Chaucer.	Grotius.	Camden.
Tasso.	Burke.	Usher, Archbishop
Jonson, Ben	Franklin.	Clarendon.
Shakspere.	Johnson, Dr. S.	Burnet, Bishop
Milton (in age).	Mackintosh, Sir J.	Buchanan.
Molière.	Walpole.	Hume.
Göethe.	Pitt.	Robertson.
Wordsworth.	Fox.	
Mrs. Hemans.	Coleridge.	
Burns.	Washington.	
	Hobbes.	

In the above instances every one is compounded with Class I, or II, or both; and would be written $\frac{I}{III}$, or $\frac{II}{III}$, or $\frac{I+II}{III}$, or $\frac{II+I}{III}$, according to the class or sub-class of profile to which it might belong.

HOBBES.

The list given is more extensive than usual; yet it might be much extended, and should comprise all the greatest names in Theology, Science, and Art.

It has been said, that "the form of the Nose in profile, decides the turn which the Cogitative power will take." Thus the Romano-Cogitative will prefer to exercise its cogitativeness in the bustle of active life, and Washington and Cromwell present remarkable proofs of the truth of this assertion.

Another striking instance is the energetic and fervent John Knox, who bearded monarchs on their thrones, and lawless nobles in their strongholds.

The major part of our illustrations being taken from purely literary men, present Greco-Cogitative Noses. It is not our intention to descant at length on persons whose works are well known, by name at least, to every one, and whose lives were, for the most part, passed in the usual monotonous tenour of those of literary men; but Bacon may be referred to as an important corroborative instance of the shrewd, wily measures by which the astute Greek prefers to further his ambition. Bacon as a man presents such a lamentable contrast to Bacon as a philosopher, and the wretched underhand means by which he attained eminence are so well known, and so painful to dwell upon, that we refrain from doing more than referring the reader to the facts for comparison with his profile. Wretchedly inconsistent as his character appears, it is not inconsistent with his Nose; and, perhaps, what are termed his inconsistencies, are only a proof that the intellectual and moral powers are distinct, and that the most profuse development of the former, cannot compensate for a deficiency of the latter.

It is unnecessary to dissertate upon the names in the present list in order to demonstrate their right to appear among the Cogitatives. No one will deny their title to that most enviable epithet, and it would be by portraits alone that the identity between their minds and noses could be exhibited to any who are incredulous on that subject. To such we can only say, examine for yourselves; the portraits are, for the most part, easily attainable; and an attentive examination of them will well repay the labour, and, without doubt, satisfy the most sceptical of the truth of the hypothesis.

The names on that list are, for the most part, names which are a volume in themselves: they write their own history; certainly no encomiums of ours can add anything to their glory. It is undeniable that it was by close cogitation, serious, hard thinking, that each of them obtained a place in the rolls of Fame;

and it is equally certain that almost every person may, by the same process obtain, if not an equal, yet certainly no mean place in the same estimable record.

It is a common and veracious observation, that certain faces prevail in certain ages; but it may be further added, that this epochal character frequently arises from the formation of the Nose, more especially of the Cogitative part.

Up to about the close of the reign of James I. the Greco-Cogitative prevailed; during the time of Charles I. and the Protectorate, the Romano-Cogitative was almost universal, and the Cogitative part was much increased in intensity. The Noses of the time are remarkably broad and thick, a circumstance which can only be attributed to the serious religious and political questions which then agitated the minds of all men. With the careless dissipated days of the second Charles came in the thin, long Greek, or Greco-Roman Nose, with little or none of the Cogitative element; and this for the most part prevailed up to the commencement of the present century. What future ages may determine to be the form of Nose characteristic of our age it is impossible to say. *We* can form no accurate judgment, for time alone can separate the tares from the wheat, and decide who are the great men of our age.

To an observant mind there is something very remarkable in the striking contrast between the physiognomies of the leaders in our own Rebellion (as it is historically termed) and of those of the French revolutionists. Besides a certain serious determination, a stern, unflinching, dogged consciousness of right, that nothing could turn to the right hand or to the left, which is visible in the countenances of the former, and to be contrasted with the flippant, wicked, blood-thirsty-looking smirk of the latter, there is a remarkable contrast in their Noses. The thick, broad, Cogitative Nose is visible in all of the former, from Old Noll himself to honest Andrew Marvel; while the void of thought, sharp, captious, *vulpine* Nose is to be seen in every one of the bloody tyrants of the French *sans-culotterie.*

The latter look like men who

> "Could smile, and murder while they smile."

The former like men who

"Put their trust in God and keep their powder dry."

Wordsworth has so splendidly and truly contrasted the men of either age that we cannot resist inserting his lines entire:

"Great men have been among us; hands that penn'd
And tongues that utter'd wisdom—better none;
The later Sydney, Marvel, Harrington,
Young Vane, and others who call'd Milton friend.
These moralists could act and comprehend;
They knew how genuine glory was put on;
Taught us how rightfully a nation shone
In splendour; what strength was, that would not bend
But in magnanimous meekness. France, 'tis strange,
Hath brought forth no such souls as we had then.
Perpetual emptiness! unceasing change!
No single volume paramount, no code,
No master-spirit, no determined road;—
But equally a want of Books and Men!

In the fifth line, "These moralists could *act* AND *comprehend*," we have a beautiful and exact paraphrase of the Romano-Cogitative, which we noticed as characteristic of the Cromwellian age—the union of physical energy with mental power

It was a remark which we heard made some thirty years ago by a very observant man, that there was a wonderful identity of expression in the countenances of all the men of the French Revolution, and that the same peculiar expression is to be seen in the faces of the conspirators of the Gunpowder Plot. Subsequent personal observation has confirmed this remark, of which it is a curious and recent corroboration, that the same expression is visible in the countenances of some of the leading Terrorists of the French Revolution.* The countenance of "bloody Mary" is an instance of the same peculiar expression.

The old gentleman who made the remark which drew our infantine attention added, (and it was this perhaps which impressed it upon our memory) that there was "*blood*" written in all their faces.

We cannot improve upon this definition,

* The physiognomy of M. Ledru-Rollin, the Communist leader, is said, by an eye-witness, to be "without one redeeming quality—insolent, conceited, reckless, headstrong, cruel."

though in one word, it might also be called a *wolfish* look—lean, cruel, hungry, grinning.

When treating of the Greek Nose, we stated that the Nose of Milton expanded into the Cogitative form when, in the latter part of his life, he was compelled to turn his thoughts anxiously and seriously to the condition of his unhappy country, and when, with a holy and unswerving determination he devoted his whole soul to the composition of a poem, whose fame should be cö-extensive with the world whose creation it described. We then claimed this instance of change of form coincident with change of character, as a proof of the correctness of the hypothesis. It was however a superfluous precaution, for the coincident change is equally true in almost every instance of the Cogitative Nose. No man can alter the profile of his Nose, but he may increase its latitudinal diameter. As to the former, he must submit to have it what shape God pleases; as to the latter, he may make it almost any shape he himself pleases—for the one indicates acquired habits, the other inherent properties.

The Cogitative Nose expands with expanding thoughts and is therefore rarely, if ever, much developed in youth; neither, on the other hand, is the very sharp or Non-cogitative Nose frequently visible in early life, for there are few to whom God has not given the elements of thought. It is our own faults, therefore, if we throw away the talents bestowed upon us, and suffer our minds to degenerate into inanity and our Noses into sharpness.

For this reason, it is a laudable ambition in a young man to cultivate a Cogitative Nose, for he can only do so by cultivating his mind. And, forasmuch as it is the only part of the Nose which is under the control of the owner, so it is that which can be most distinctly judged of and its expansion watched; for, though the owner can never see the perfect profile of his Nose, he may always form a correct estimate of its *breadth*. We should be quite justified in adding this to the numerous proofs of design in the adaptation of the human body to the soul, but as many persons cannot surmount a certain sense of the ridiculous in the subject before us, we forbear. Those who are impressed with the truth of our system will at once admit the inference, and perceive its value in Natural Theology.*

As it has been deemed unnecessary to extend the present chapter with any biographical or critical sketches of the examples adduced in corroboration of Class III., we will devote the next to the more useful task of inquiring how a Cogitative Mind and its certain accompaniment, a Cogitative Nose, may be acquired.

* We trust no one will misunderstand these observations, but give us credit for making them sincerely and with all reverence; firmly convinced as we are, that if the system is true, it *must*, like all other sciences furnish its quota of proofs of design in the universe.

CHAPTER V.

HOW TO GET A COGITATIVE NOSE.

It is a great and prevalent mistake to imagine that a Cogitative mind (and Nose) is to be acquired by reading alone. It is almost certain that, as books multiply, Cogitative Minds decrease, for how is a man to think, if all his thinking is done for him? The mind, when constantly supplied with extraneous thoughts must, without great care, lose the habit of generating internal ones. All the greatest thinkers have been the first in their department of thought. Homer, Dante, Chaucer, Shakspere, Bacon, &c. These men, as compared with even mediocre men in our day, had very little learning,—but they had vast wisdom.

Read Bacon's Novum Organum and Sylva for instance, and see how few facts there are in them but such as are either now known to, or laughed at, by every school-boy; yet direct your attention to the train of thought, to the generalizations from these simple facts, to the originality of the deductions, and behold how the dwarf in Knowledge becomes a giant in Wisdom! It is even true that Bacon was behind his cotemporaries in many matters of mere knowledge; yet the majesty of his wisdom was so vast that it still rules, and ever must rule, the world of science.

So, as on the one hand, a man may have wisdom and yet want knowledge; on the other, he may have all knowledge and be able to discourse of all things, from the hyssop to the cedar, and yet want wisdom. It is of no use to read and accumulate facts if we do not also *think*. Better indeed to think and never read, than read and not think. If a man does not think for himself, if he does not originate ideas, if books are not to him *only* the elements of thought, if he is not fully and immoveably impressed with the conviction that two and two make five, or any greater number which the Cogitative Mind can evolve, he has no chance of becoming a wise man, whatever his learning, and however profound his acquaintance with the thoughts of other men.

But you reply, two and two do not, and cannot, make five, &c. We rejoin, they as certainly and unquestionably do in metaphysics, as they certainly and unquestionably do not in physics. True, in physics, two and two things, two and two facts make four, and only four; but if the mind, when in possession of those four, can generate nothing more from them it is a hopeless case with that mind. If, upon the recipience of such four facts the mind remains contented with the mathematical fact that, from four units it has segregated four, it is, and for ever will, remain stationary; it has gained nothing, and might as well have left those four facts in their original units, for their addition has not added to it one particle of wisdom.

Facts are, or ought to be, only the generators

of ideas. Facts in themselves are utterly worthless; it is in their associations, in their consequences, their bearings on each other; it is as they support or refute systems, theories and other mind-born facts, that they are of value. Now, it is only by the action of mind upon them that they have associations, consequences, &c. Without mind, facts must for ever remain units; even though added together, *ad infinitum*, they have no natural co-unity, no cohesion, no affinity for each other. A thousand facts added together are still but a thousand units, unless mind has cohered them into a system. This done, you clearly have the thousand facts still, but you have *also* something infinitely more valuable, you have a mind-born fact, a deduction, a system, hypothesis, theory, axiom, or whatever you please to call it.

Cordially as we hate coining new words, we still more cordially hate the German fashion of hooking two words together by a hyphen and calling the junction an addition to the language. But we are compelled, in order to save circumlocution, to coin a word to express those facts which spring from Mind, whether, as in moral philosophy, purely metaphysical, or as in natural philosophy, generated by Mind from Matter, by Reason from Experience. Such facts we would beg to call noögenisms (νοος, *mens*, *cogitatio*, and γενος, *natus*, *progenies*); therein including all mental offsprings or deductions, whether called hypotheses, theories, systems, sciences, axioms, aphorisms, &c.

Noögenisms, therefore, are those facts which mind generates from other facts without annihilating the latter; hence it is said that, metaphysically, two and two make five. Thus, mind, contemplating the physical facts of the super-position of strata, deduces from and adds to them this metaphysical fact or noögenism: —Strata were deposited successively.

Herein appears too an essential difference between Mind and Matter. If diverse substances, having a natural affinity, be amalgamated, a new substance is obtained, but the elements are lost. Of hydrogen and oxygen water may be made, but the gases are forthwith lost in the fluid; the procedure may be reversed and the water be converted into gases, but the water has disappeared. This is not so with mind and noögenisms; for however closely, by a mental synthesis, diverse facts may be united into a new fact or noögenism, the latter is obtained without losing the former or elementary facts, which remain as knowledge, elements of wisdom to support the noögenism or create others.

We see then, that while Mind is crescive Matter is not. Matter is neither crescive nor decrescive. It may be changed into divers forms, animal, vegetable, or mineral, but it never can be varied in *quantity*. The six feet of animated clay dies, it rots in the silent tomb; years pass by. The hand of affection which protected the loathsome, yet—for the once animating spirit's sake—beloved, remains is cold and rotted too. The sepulchre so long forgotten and deserted again becomes of interest to the brother of the hyæna, and the resurrectionist —the antiquarian. He, in his cool business-like phraseology, opens a barrow or exhumes a tomb, and finds—what? A pound of dust! The sole visible remains of a gigantic hero or a stalwart king. Yet, is not one particle of that ancient demigod perished. Every atom is, in some shape or other, in the universe. Some atoms may "have gone a passage through the guts of a beggar," and so have nurtured another human form; some may have stopped a beer-barrel and so

> "Imperious Cæsar dead, and turn'd to clay,
> Might stop a hole to keep the wind away."

The theory of the metempsychosis is true of Matter; and as the ancient sages believed the soul to be material, that theory, so far from being violently absurd (as we in the pride of better knowledge are apt to term it), was almost the only theory which the thinking and observant mind could of itself elaborate. Hence the adoption of that system by far-distant nations is no proof of inter-communication. What the Brahmin in India found a natural result of the doctrine of the materiality of the soul and its consequent analogy to everything else material, the Druid in Britain would arrive at with equal ease.

But Mind is both crescive and decrescive; and it is another peculiar property of Mind, that it is never stationary,—it is always changing, increasing or decreasing. This is an important consideration; a fearful responsibility cast upon it. If the one talent (and God has so benignly ordered it, that no sane, and therefore responsible, mind is devoid of, at least, one talent,) is hid in a napkin, the servant is condemned and his talent taken from him. But if the talent is put out to use, it will increase and grow, and make other talents, and the lord of that servant will receive his own again with usury. For, having endowed man with this crescive power, He justly demands that power to be exercised and the mind to be enlarged and expanded "by every one according to his several ability," so that He may reap the harvest which His well-rewarded servants have gathered in, "reaping where He hath not sown, and gathering where He hath not strewed."

The very cause of this crescive power of mind is, that the sum of the units aggregated by mind is greater than the mathematical sum of the units; and the cause of this is, that facts, the elements of noögenisms, are not, like chemical elements, lost in the fact compounded from them, but retain likewise a separate independent existence, capable of being again compounded into other noögenisms, and still ever without losing their original forms.

It will now be understood what is meant by two and two making five, &c.; and until a man is incontrovertibly convinced of the possibility of this he will in vain multiply facts. Facts must be added together, not for their arithmetical product, which is Knowledge, but for their metaphysical product, which is Wisdom. You will frequently hear asked by utilitarians, what is the use (*cui bono?*) of such and such knowledge? Remember that the use of all Knowledge is to feed the mind and to generate Wisdom, and you will always have this ready and sufficient reply, "It is food for thought."

And here it may not be out of place to endeavour to point out by an example the difference between knowledge and wisdom, and at the same time elucidate more clearly how the former is to be made subservient to and the genetrix of the latter. We observe that a certain quartz-stone is round. We have learnt two facts, the nature and form of the stone. Now what is the value of those facts *per se?* The recipience of them has increased our *knowledge*, but is the mind strengthened or rendered one jot *wiser?* We trow not. But as a key or foundation to an aqueous *theory* of geology they are almost infinitely important. The Cogitative Mind perceives that the round stone must have once been an angular fragment broken off from some rock of quartz, and asks, "How came it broken off? and how came it round?" The answers are a whole system of geology; nay, perhaps an entire system of the universe, a noögenism of the sublimest kind.* Have not these facts generated? Is it not clear that, if the physical

* The use of this word would often save the quibble, whether a system is entitled to be called a science or only a theory or hypothesis. Thus both the advocates and the opponents of phrenology or geology might agree to call them noögenisms. For this reason we apply the word here to geology, which some persons assert to be more than a mere hypothesis, while others deny its claim to be called a science. At present we claim for Nasology no higher title than that of a mental deduction from facts or noögenism.

units had remained metaphysical units they would have been valueless? but being submitted to the powerful energy of intense thought they become the parents of a noögenism, into which "the angels desired to look," and at the first dawn of which, from the primæval chaos, "the morning stars sang together, and all the sons of God shouted for joy."

Neither is this instance fanciful; for, while we write, it reminds us that this identical simple fact,—a round pebble on a common,—appeared to Paley to be one from which the mind could evolve nothing, and therefore he contrasted it with a watch, whose mechanism led the mind to theorize on its causes and origin; whereas, a recent commentator thereon justly observes, that the stone was as fertile a source of cogitation and as able a guide "from Nature up to Nature's God," as the watch was from itself to *its* maker.

From this example let us take warning, that facts be not to us nothing more than round stones. Let us be careful never to let our minds rest content with the mere accumulation of facts, but ever strive to build them up into something more useful and ennobling. Let us use them as bricks—mere logs of burnt clay in themselves, but fit to build glorious monuments of the sublime power of human invention. Let us remember that Ideas are the only things of real permanent value in this world; and that though we may store our brains with Facts till our heads burst, unless those facts are to us only generators of Ideas, we have not, and cannot acquire a Cogitative mind; we may have Knowledge, but we have not Wisdom. A wise man hath wisely said, that "the wise man is" —not he who knoweth *things*, but—"he who knoweth the *interpretation* of a thing" [*Eccles.* viii. 21]; and for this purpose only it is that, "Wise men lay up knowledge" [*Prov.* x. 14], for "Wisdom finds out knowledge of witty inventions." [*Prov.* viii. 12].

In order to effectually discipline the mind to attentive study and to save it from the strong temptation which is offered to desultory reading, it is advisable for the adult and partially educated student to form an hypothesis and read up to it. To reverse, in fact, the Baconian principles of philosophy, and to study from hypothesis to facts, and not from facts to hypothesis. This is, it is true, opposed to modern philosophical principles, but properly modified and carefully guarded against self-conceit and dogmatism it is almost the only proper and effective mode of study. It is the ancient or Aristotelian mode; and though, when refuted by Bacon, as a mode of "discovering the sciences" it had become shamefully abused and degenerate, it has produced more great original thinkers than the modern. Observe, that we recommend it only as a mode of *study*, *i. e.*, of disciplining and exercising the mind, for beyond the purpose of *training* it should not be pursued. It is too dangerous to be prosecuted far, for the mind which has long formed and nursed up a favourite hypothesis is unwilling to abandon it, and is too apt to force all facts into accordance with it, instead of modifying or abandoning it as new facts arise.

But the great advantages of this plan, as a training process, would appear to be—1st. That the mind being thus occupied with an hypothesis has always that to direct its researches in a settled, uniform, and definite course. 2nd. That every new fact accumulated is immediately compared with the hypothesis, and is incorporated or written off as *contra*, after this mental exercise, as occasion may require. Thus no fact ever comes into the mind without being subjected to thought and giving exercise to the important faculty of comparison. And this process of comparing, to which every fact must be subjected, will not only impress the fact and its comparatives on the memory, but will powerfully tend to exercise and strengthen the Cogitative powers; for there is no operation of mind which more actively calls into energy all the faculties at once than comparing, because to compare two things fairly we must (so to speak) know the length, breadth, depth, density and powers of each. 3rd. A steady habit of reading is acquired; we read with a definite aim—the establishment or refutation (we ought not to care which) of our hypothesis, and, however wide and discursive our reading, there is little danger of its becoming desultory—that curse and bane of modern mind. The Baconian process of accumulating facts before hypothesising, almost demands desultory reading, for the mind sees no fixed end towards which it shall arrive; it is not permitted to guess what may be the result of its studies, and hence too often loses all interest in them, and remains content with the barren accumulation of things.

What we would suggest may be thus illustrated: Let a man, intending to study history, first adopt an hypothesis—of course he must have some pre-knowledge. It matters not what the hypothesis, so that it is likely to involve a very wide field of inquiry. If he contemplate primæval history, let him adopt some such proposition as this, "Whether we can infer from the institutions of mankind that they all spring from one common ancestor?" Or this, "Whether any nation whose national records have been preserved were the first owners of the soil?"

Is it not obvious, that with some such proposition before the mind it will take much more interest in and more steadily direct its studies, and that facts will be more easily remembered, from their bearing on the hypothesis, than if merely received as naked, isolated units?

The only precautions to be taken are, not to be too strongly wedded to either side of our hypothesis, nor to sit down too soon satisfied that it is proved or disproved, nor to set up for teachers and discoverers, while we are only learners and making discoveries.

It will be seen hereafter that, notwithstanding what has been said, we differ not at all from Bacon himself; we differ only from his pseudo-disciples, who have no more in common with his enlarged views of the uses of science than the schools had with Aristotle, or the New Academy with Plato. Nevertheless, we well know that we shall be well abused by these disciples as an impugner of Bacon, and as a heretic to his philosophy, just as your pious people condemn as an infidel or atheist every one who denies any dogma which their wild enthusiasm has grafted on the Bible. It is not in religion alone that bigotry is to be found.

Bacon himself pursued the mode of study which we suggest. At fifteen he formed an hypothesis, and devoted his whole life to its elucidation. The hypothesis round which, as a centre, he gathered every fact within his reach was this: Whether or not the Aristotelian was the best mode of cultivating the mind, and of discovering the sciences?

He seems at first to have been disposed to think that it was neither; but the conclusion to which he finally came, after many years of close thought and arduous study, was, that it was the best mode of cultivating the mind, but the worst mode of discovering the sciences. He did not soon sit down satisfied that he was right, and set up for a dogmatic teacher of his new philosophy. He waited patiently for any new light which years and experience might throw upon it, either bringing out more brightly its beauties or disclosing more satisfactorily its errors. Once in each year he reviewed it and tested it by the new facts which he had gleaned during the year's studies. Once in each year, for twelve long years, he wrote out with his own hand, altering, condensing and verifying his *Novum Organum* before he published it.

So much stress has, notwithstanding this

illustrious example of the master, been laid, ever since the publication of the Baconian or inductive philosophy, upon the bare accumulation of facts, and so much has been written against generalizing and hypothesizing, that it may be as well, before quitting the subject, to point out wherein the disciples of Bacon have neglected the precepts of their master; and to inquire whether this neglect, and the only *partial* adoption of his teachings, have not contributed greatly to the advancement of mere Knowledge at the expense of true Wisdom, and thus been a very important cause of the degeneracy of modern mind.

Bacon seems to have foreseen this effect of the exclusive adoption of the experimental part of his philosophy—the only part which men have yet had the courage to adopt—when he said, "Our way of *discovering the sciences* almost *levels the capacities* of men, and leaves little room for excellence, as it performs all things by sure rules and demonstrations, and therefore these discoveries of ours are, as we have often said, rather owing to felicity than to any great talent, and are rather the production of time than of genius."* It was for this reason that he so earnestly, as we shall see hereafter, insisted against its use by young and common minds, or as a means of mental cultivation. And too truly has the prophetic caution been fulfilled! Nevertheless, as it will be loudly denied that modern mind is degenerate, it may be as well to ask how much we are in anything, except physical science (facts, or what Bacon calls, "Experience"), in advance of our two-hundred-years' dead ancestors. Array the names in our list of Cogitatives, chronologically and analytically, or do so by any list of great thinkers, and you will scarcely find a proportion of one since 1700, to three who lived between 1550 and that date.

Nevertheless, though there is this falling off in Wisdom, how vast has been the accession of Knowledge. Bacon, in his day, complained that the former (Reason) had gone on without the latter (Experience); so that, while mind had attained the highest flights of which it seemed capable, the arcana of nature were yet unexplored,

* Nov. Org., Sec. VII.

and little or nothing had been done to advance man's physical welfare. He said that, hitherto, reason and experience were as new gifts of the gods:—the one laid on the back of a light bird, the other on a dull ass, and that as yet they had not been united. His object was to unite them; to this purpose, he devoted his gifted mind and strained his utmost energies. Yet if he were living now he would be compelled to make the same complaint, with this variation however, that men have abandoned the burden of the bird, and have loaded themselves with that of the ass.

While then we admit the rapid advancement of knowledge, let us pause a moment and inquire if it is not a proof of the degeneracy of mind and the decay of wisdom, that, in that which is purely mental or dependent on mind, we have no names of equal note with the names of those who lived before the exclusive adoption of the experimental part of the Baconian philosophy. Where is the name in poetry to set against Shakspere and Milton; in metaphysics to match with Locke, Hobbes, &c.; in deduction from facts and generalization with Bacon, Newton, Halley, &c.; in theology, with the hundreds of names which yet eclipse all modern commentators? It may perhaps be said in reply, if we have not such great minds, we have a larger number of thinkers of lesser magnitude. This is doubtful. Time has obliterated the swarms of lesser fry who, like their congeners of the nineteenth century, lived their day and gained a temporary fame in the sixteenth and seventeenth centuries. But further, it is easy to be a triton among minnows. It is as easy now-a-days to set up for a literary character and "write a book" without an idea, as it is for an insolvent man to pass for a rich one and live sumptuously on borrowed capital and paper money. Our thousands of authors are but the minnows which sport in the shallow brooks and live their little day in glorious self-gratulation on the laudations of their brother minnows; but if they happen to get out into the deep, strong waters and a triton turns his stern eye upon them—pop—they turn their tails round, dive to the bottom and are seen no more. Thus it was with our novelists; they shone and blazed away—happy, glorious book-wrights—till the triton Scott came athwart their path, and straightway they were gone. And surely, surely, we now again want another Scott to demolish the rapidly increasing tribe of cachinnators, who appear to deem that the proper end of light literature is just to raise a temporary laugh and be forgotten. Heaven send us salvation from more Jerrolds, à Beckets and the whole tribe of ephemeral laughing-stocks! It is the same in other and more important departments of literature. Our historians are mere compilers of old letters; we fly to Germany for historical criticism and acute generalizations from facts, contenting ourselves with laboriously picking up a few obscure facts for the use of our more deeply-thinking neighbours; who are treading in the paths which our own sages trod two hundred and fifty years ago, because they have not yet placed the exact sciences at the head of intellectual pursuits, and abandoned thought for mechanisms, generalizations from facts for the barren accumulation of facts.

The complaint of Lord Bacon is truer now than it was in his time: "If a man turn his eyes to libraries, he may perhaps be surprised at the immense variety of books he finds; but upon examining and diligently weighing their matters and contents, he will be struck with amazement on the other side; and after finding no end of repetitions, but that men continually treat and speak the same things over and over again, fall from admiration of the variety into a wonder at the want and scantiness of those things which have hitherto detained and possessed the minds of men." Unhappily his system, by the universal and indiscriminate adoption of only its lower and material offices to the exclusion of those higher ends which he contemplated from it, and by its being used as a mode of cultivating the mind, as well as a means of discovering the sciences, has rather strengthened than weakened the justice of these censures. Our Augustan age of thought is still that of Elizabeth and James I.; the latter part of the sixteenth, and the early part of the seventeenth centuries still outshine the nineteenth in loftiness of thought and solidity of learning; yet we complacently boast of our progress, because we rattle through the fields of learning at ten times the speed of our ancestors, as we do over our railway-sected country, gleaning about as little information of the one as of the other. We dash through the deep cuttings and dark tunnels of literature at railway speed, taking assertions for facts, and empty declamation and tawdry immorality for sense and religion; and then, like the nervous lady who rides through a railway tunnel without fainting, congratulate ourselves on having accomplished some gigantic feat; though we have learnt just as much about the subject of our studies as she has of the construction of the tunnel; but having, like her, fretted and fumed for a few minutes at some dark difficulty, we unite with her in thinking ourselves very valiant and clever people.

We avail ourselves of the roads and paths which others have made, and never stop to examine their solidity or foundations, or the principles on which they are constructed. We lose the habit of deep investigation and close thinking by a long and entire reliance on others, and our minds become dissipated, and a prey to all the silly novelties which spring like ephemera from the almost stagnant pools of modern brains.

This mental dissipation and its concomitant evil, reading for the purpose of killing time,—with far more baneful effects than never reading at all, but relying merely on our own serious excogitations,—are curses from which we ought earnestly to endeavour to save ourselves. This we can only do by sternly exercising the mind in settled definite habits of thought, by placing before it a determinate aim and end to its cogitations. It must know beforehand whither it is tending, so that, as it proceeds, it may note its progress, and be able to judge whether it is advancing or receding. It would be as absurd for a man to start on a journey without knowing whither he was going, but to be continually trying first one road and then another, in hopes it would bring him somewhere, as it is for a student to sit down to study without any definite purpose or view before him. True, the traveller might pick up many facts and get some knowledge in his desultory course, and so might the student; but neither would be advanced on his journey or have gained any true wisdom. Yet this is the course of modern study. Loose desultory reading: a vague acquisition of unconnected facts is alone aimed at. Witness the transactions of our scientific bodies—a huge undigested mass of valuable facts; the raw materials, the bricks of knowledge, which no one has dared yet to generalize or build up into a harmonious and well-proportioned temple of wisdom.

Modern *savans* shrink from using the materials which, for several centuries, thousands of laborious literary ants have been collecting. Like the unhappy Psyche, doomed by the inexorable Venus to arrange and sort into respective heaps a confused mass of wheat, barley, rye, millet and other kinds of grain, they sit down in despair of accomplishing the apparently hopeless task. Frightened at the gigantic labour, they not only fly from it themselves but condemn every one who attempts to arrange systematically the grains which, assorted, would afford valuable seed for fresh crops of food, but which, while thus intermingled, are utterly useless and unproductive. With an insane determination *not* to see the work which it is the duty imposed on the soul (Psyche) by the prolific powers of nature (Venus) to accomplish, they go on adding to the heterogeneous heap, and endeavour, by loud and clamorous applauses of those who are mere collectors like themselves, to drown the voice of those who would incite them to the enjoined and higher duty of assorting and arranging.

Should any one, like the able but mistaken author of the "Vestiges of the Natural History of Creation," endeavour to bring Thought to bear upon these dry bones and make them live, to generalize and build up a system from them, great is the outcry and terrible are the denunciations. The modern Prometheus, who would animate with the celestial fire of forethought the clay which lies a dead and useless mass at his feet, is clamorously damned by his timid brethren the Epimethei, the after or past-thinkers; and, unless he is endowed with more than mortal power, he must submit to have his heart daily devoured by the racking fiends—envy, hatred, malice and all uncharitableness.

One of the laborious ants of whom we have been speaking asks, "For what do we read?"

and complacently answers, "To know *facts.*"* Indeed! The highest office of mind is to make itself a barren store-house! To us it appears, on the contrary, that we should read and study generally, not to *know* facts, but to *be wise* from facts, to make the head wiser and the heart better. The mind is not to be considered as a mere granary and barren receptacle of literary food, but rather as the stomach which converts into a new substance—assimilating good healthy flesh and blood—the heterogeneous materials which are put into it. Another ant, of no mean pretensions among his brethren, enthusiastically endeavours, by promises of "literary glory," to incite some of them to pile up into one heap the confused materials they have collected. "Let us see," he says with childish glee, "how much we've got! You John, and you Willy, and you Bobby bring what you've collected! There pile it up! Make a snow man; cut him eyes, and

* How different is the language of the disciple from that of the master! Bacon himself says, "Read not to contradict nor to believe (*i. e.* for facts), but to *weigh and consider.*"

nose, and mouth." There he is with a pipe in his white lips. Doesn't he look sage, and grave and solemn? Dance round him, ye children; clap your hands and be merry. Rejoice over your work while it lasts. The first warm breath of spring will melt it away. It is no man, it has no life, it is cold and dead. The snow, give it what shape you will, is snow still. You have collected much, but you have got nothing new out of your collection. But lest it should be supposed that we belie this celebrated ant—this collector of grain—we will quote his own words: "Within the last two hundred years (says Professor Playfair), or since Galileo and Bacon taught us this great lesson, *we have been employed in recording facts* in ten thousand several volumes. But thus scattered, they lose so much of their value and importance, that, in another age, we may hope some aspirant after literary glory will perform the Herculean labour of condensing the whole into (What?—a system of the universe? a better knowledge of nature? No!) *a volume!*" A *volume!* that is to say, gather the scattered masses into one heap as heterogeneous as the scattered masses; pile up the snow, strewed over pathway and field, hedge and ditch, into a snow man. That is the highest aspiration of this Professor of divers learned societies. His grovelling soul soars not to the hope that any new fact may be extracted by mind from this vast heap of raw materials. He knows not that, metaphysically, two and two make five, and that without any other material additions, without any more ant-collections, the heap may be made to grow and swell, that the spirit of life may be breathed into it, and that, wedded to mind, it may even become the prolific parent of new facts of a far higher and more enduring nature than any in his boasted volume. Facts, which, having mind for one of their parents, will with filial love pay back in tenfold blessings the life given them; facts which will lead that parent to unravel the mysterious secrets of nature and enable her to behold the wonderful arcana of its Holy of Holies.

This is the purpose for which we should read, and this the glorious end for which we should collect facts; instead of merely contenting ourselves with being employed, as Playfair too truly says we have been since the time of Bacon, "in recording facts in ten thousand several volumes," with no higher aspiration than that some laborious stable-cleaner may sweep them up, hay and straw, corn and rubbish into one vast heap. Since this was written, it pleases us to see that the able author of "Vestiges of the Natural History of Creation" has in his "Explanations" spoken to the same effect and added another instance of the low estimate formed by modern scientific minds of the uses of facts. "From year to year and from age to age we see scientific men at work, adding, no doubt, much to the known, and advancing many important interests, but at the same time doing little for the establishment of comprehensive views of nature. Experiments, in however narrow a walk, facts, of whatever minuteness, make reputations in scientific societies. All beyond is regarded with suspicion and distrust. The consequence is, that *philosophy*, as it exists among us, *does nothing to raise its votaries above the common ideas of their time*. Let me call upon the reader to bring to his remembrance the impressions which have been usually made upon him by the transactions of learned societies, and the pursuits of individual men of science. Did he not always feel that while there was laudable industry and zeal there was also an intellectual timidity, rendering all the results philosophically barren?

"Perhaps a more lively illustration of their deficiency in the *life and soul of nature-seeking* could not be presented than in the view which Sir John Herschel gives of the uses of science, in a Treatise reputed as one of the most philosophical ever produced in our country. These uses, according to the learned knight, are strictly material—it might be said sordid—namely, 'to show us how to avoid attempting impossibilities, to secure us from important mistakes, in attempting what is in itself possible by means either inadequate or actually opposed to the end in view; to enable us to accomplish our ends in the easiest, shortest and most economical and most effectual manner; to induce us to attempt and enable us to accomplish, objects which, but for such knowledge, we should never have thought of undertaking.'

"Such results, it will be felt, may occasionally

be of importance in saving a country gentleman from a hopeless mining speculation, or in adding to the powers and profits of an iron-foundry or a cotton-mill, but nothing more. When the awakened and craving mind asks what science can do for us in explaining the great ends of the author of nature, and our relations to Him, to good and evil, to life and to eternity, the man of science turns to his collection of shells or butterflies, to his electric machine or his retort, and is mute as a child who, sporting on the beach, is asked what lands lie beyond the great ocean which stretches before him."*

This is unhappily too true a picture of modern science. Every effort is made in scientific works to impress the material and sordid money-getting uses of science as its only true end, and the highest relation which it bears to humanity. Read any tract on the uses of geology, and is there a word of high hope that the addition which recent discoveries in this department have made to knowledge will assist

* Explanations, 2nd Edit. p. 78.

in raising and elevating the mind, or throw any new light upon the mysteries of nature?

Not a word: but it is carefully detailed how an acquaintance with the order of stratified rocks will facilitate the discovery of minerals, or the boring of Artesian wells.

Are the uses of astronomy dwelt upon, we are taught that it enables the seaman to navigate trackless seas for commerce or for war. Are the purposes of chemistry detailed, we learn that it is fertile in assisting the manufacturer to cheapen his goods, and undersell his less experienced neighbour.

And are we to believe that for these base uses it will be given to man to penetrate the wonders of the universe, and read the unexplained mystery of its creation? Surely not. No, verily, we must raise our souls far above these debasing cares, before the great and beneficent God will permit us to understand His sublime works. We must come to the task with clean hands, with pure, holy, unsullied minds, with humble, but high aspirations, with the submission of little children, but with the elevation of pure wisdom

Is it possible that mind can progress at all, if it is for ever fixed on the earth, grovelling after barren facts and never lifted up to heaven, nor exercised in contemplation of the discoveries it has accumulated? Is it possible that the mind can ever be wise which believes that it must study for facts, and not to 'weigh and consider?' Must not the former for ever remain the mere basket of the rag-and-bone collector? the receptacle and dead vehicle of material things? Is it not better that the mind should be exercised, like 'a light bird,' in the wildest and most visionary dreams, than be reduced to such a 'dull ass,' or dead entity? If the student would avoid the latter, he must abandon the mere accumulation of facts for the comparing and weighing of evidence, the calm looking for results, and the deliberate generalization from the facts collected by the fact-collectors; the rag-and-bone-pickers, the hewers of wood and drawers of water of the human race. Nevertheless despise them not; they fill their allotted station in the world; they are as necessary to the thinkers as the different ranks in society are to each other. Bear in mind that

Bacon never intended his system for *students*, or to be used as a mental exercise. He only proposed it as a means, (and confessedly the only true means) of 'discovering the sciences,' and not as a mode of 'cultivating' the mind. It was to be the exercise of the experienced and completely cultivated mind only, 'of the man of riper years, sound in his senses, and of a clear, unbiassed mind.'* He foresaw and cautioned against its abuse by 'vulgar minds.' And in the sense used by him all young and learning minds are vulgar (common) minds. The specialities which must distinguish them from the common herd, are as yet unknown and hidden beneath the crust of inexperienced ignorance.

He himself earnestly prays that his own and the Aristotelian system may live together, and go hand in hand, the latter to cultivate the mind, the former to discover facts. His words so long forgotten and unheeded by his disciples are: "Let there be therefore, by joint consent, two fountains, or dispensations of doctrine, and

* Nov. Organ.

two tribes of Philosophers, by no means enemies or strangers, but confederates and mutual auxiliaries to each other; and let there *be one method of* CULTIVATING, and *another of* DISCOVERING the sciences. Nor is our's very obvious, *and to be taken at once*, nor tempting to the understanding, nor *suited to vulgar capacities*, but solely rests upon its utility and effects" (*i. e.* upon the way in which it is used and the results which proceed from it). "But no one, sure, can suspect, that we desire to destroy and demolish the philosophy, the arts, and the sciences at present in use; for, *on the contrary, we embrace their use*, and willingly pay them all due honour and observance. For we openly declare that the things we offer, are not very conducive to these purposes (mental exercises), as they cannot be brought down to vulgar capacities, otherwise than by effects and works."* Therefore in advocating the retention of the Aristotelian mode of thinking for students, we do but follow in the footsteps of his great opponent; who yet opposed only

* Nov. Organ. Pt. 1. Sect. 7.

when that ancient philosophy was carried beyond, and out of its proper department—the cultivation of the powers of thought, into the discovery of the sciences.

"The two faculties of reason and experience," says Bacon, "should be properly joined and coupled together." Reason without experience (facts) he compares to a light bird; Experience, without reason to a dull ass. It is better to be the bird than the ass; it is best to be neither, and yet both. It is only by joining experience with reason that the 'sober certainty' of the quadruped can be coupled with the 'waking bliss,' the ecstatic heavenward flight, of the light and joyous bird. If, like Bacon, we were to endeavour to read the fable of the Sphinx, we would say that it represents the wise mind, which has united reason and experience into a beautiful form; comprehendible by man, but most hard to be comprehended. Its human head portrays that to intellectual man alone it is given to join together its other forms, the wings of a bird, reason; and the body of a quadruped, experience. It is beautiful, for such union is the perfection of wisdom, and 'O how comely is wisdom!' It is cruel, for many lives must be sacrificed 'ere it can be discovered, or the problem of its nature be solved.

Far different from the master himself, who saw in his philosophy the attainment of high and holy purposes, his pseudo-disciples shrink not from avowing that the material uses of philosophy are of higher import than the metaphysical. And it is because writers of no mean powers have, while setting themselves up as encomiasts and expounders of Baconism, utterly lost sight of the higher and godlike purposes which Bacon hoped to see his system promote, and have exalted only the simply mean and sordid uses, which, as tending to man's temporal comforts, Bacon's large heart also desired to increase, that we have so far enlarged our observations hereon; and shall 'ere we conclude, set a few extracts from these modern views of Baconism in opposition to those of Bacon himself. From these we shall see that, with regard to their views of the objects of philosophy, no two systems can be more opposed than that of Bacon himself, and that of the modern utilitarians, who dare to dub themselves his disciples. The latter seek in science nothing higher than base utilitarianism, thus elevating the body at the expense of the soul; the former sought utilitarianism in company with the attainment of pure truth and the investigation of the hidden secrets of nature, thus elevating both soul and body.

It was the fault of the ancient philosophy that it endeavoured to elevate the soul at the expense of the body, and to separate that which God has joined together; it is equally the fault —but a far more baneful one—of modern utilitarianism that it endeavours to elevate the body above the soul, and treats the comfort of the former, as of far higher importance than the exaltation of the latter.

Bacon alone, truly wise, sought the well-being of both; and he alone pointed out that the well-being of both lay in the same path, and might be prosecuted simultaneously. While the ancient philosophy feared to defile the soul by contact with what was falsely called the base in nature, and the utilitarian dreads to have his sordid soul elevated above the same operations,—which he equally terms base, yet loves to degrade himself to—Bacon acknowledged nothing base in nature, and feared not to study her simplest and meanest operations in the pursuit of truth. He knew that whatever advances the soul makes in knowledge and wisdom, must be made through, and by means of the body; therefore, the latter was not to be despised, but by all possible ways and means to be made the efficient hand-maid of the former. He knew that though the eye sees not, and the ear hears not, yet that the soul, in this mortal state, could neither see nor hear without them, and that by increasing their fact-transmitting powers, he was developing the fact-generating powers of the mind.

It was for this reason that he contemned not to give his mind to experience, to making telescopes and ear-trumpets; but nevertheless he did not regard them as the ultimate and sole end and aim of his philosophy. His views of the ends of philosophy were, as we shall presently see, to the full as high and lofty as those of Plato and the Grecian philosophers;

he only sought to arrive at those ends by means different to those which they pursued. They both sought the same objects—Truth, and the discovery of the secrets of nature; but while the one foolishly did this by opposing nature, and acting in contradiction to her mandates, the other did it by following her patiently through all her devious windings.

The modern Baconian school of utilitarians err in stopping half-way, and in mistaking what Bacon merely deemed media, for the ultimate ends of his philosophy. Whirled along by a steam-engine, informed by a telegraph, freed from pain by chloroform, the utilitarian deems such-like products of the inductive philosophy, to be the *summum bonum* of its founder; forgetful that he considered such to be but the means to a higher end, and has said that "the *summum bonum* of human nature is the possession of truth, for this is a heaven upon earth."

But the better to understand this, let us contrast modern Baconism with Bacon,—"ab uno disce omnes."

Mr. T. B. Macaulay, a masterly and deservedly popular writer, has undertaken to give a more correct analysis of the objects of Baconism than is usually entertained; but as it happens to be only an analysis of modern utilitarianism, we will avail ourselves of it as a contrast with Bacon's own aspirations of the benefits to be derived from his system.*

Hear the utilitarian's version of Baconism in contrast with the ancient philosophy.

"Plato, after speaking slightly of the convenience of arithmetic in the ordinary transactions of life, as to make men shop-keepers or pedlars, passes to what he considers as a far more important advantage. It habituates the mind, he tells us, to the contemplation of *pure truth,* and raises us above the material universe; and he advises his disciples to this study, in order that they may learn to fix their minds on the *immutable essences of things.* Bacon on the other hand, valued this branch of knowledge

* Historical and Critical Essays, vol. II. The reader who wishes to form an estimate of the sordid views of the utilitarian school had better peruse the whole of Macaulay's Essay on Bacon.

only on account of its uses with reference to the visible and tangible world.

"Of mathematics, Plato says the real use is to lead men to the knowledge of abstract *essential, eternal truth.* Bacon valued mathematics chiefly, if not solely, on account of those uses which Plato deemed so base—its application to mechanics, &c. If Bacon erred here, we must acknowledge that we greatly prefer his error to the opposite error of Plato.

"To sum up the whole," says this eulogist of what he deems Baconism against the ancient philosophy as explained by Plato, "we should say that the aim of the Platonic philosophy was to exalt man into a god. The aim of the Baconian philosophy, was to provide man with what he requires, while he continues to be man, and to supply his vulgar wants. The former aim was noble; but the latter was attainable. Plato drew a good bow, but he aimed at the stars; therefore the shot was thrown away. Bacon fixed his eye on a mark, which was placed on *the earth,* and within bow-shot, and hit it in the white."*

* Essays, vol. ii. p. 386—403.

If this were a true picture of Bacon's mind, how sad, and low, and grovelling, must it have been. Accustomed to grieve that he suffered his soul to be polluted by contact with the world, and bowed his heart beneath the love of ill-gotten gold, we have yet found consolation in the thought that the man and the philosopher were two; and that we might dwell with rapture on the latter, take him to our heart, and make him our mind's companion without defiling ourselves with the former. But if this were a true picture of the philosopher, we must turn from him with disgust, as one whose soul was so imbued with the low and sordid, that no intellectual powers, how sublime soever, could elevate it above what was low and sordid, mean, and base.

Sick at heart and disgusted with humanity we must turn with joy to him who sought 'to exalt man into a god,' who urged us 'to the contemplation of pure truth,' 'to fix our minds on the immutable essences of things,' and 'the knowledge of the abstract, essential, eternal truth.'

But thank God, it is not a true picture of

Bacon's mind and purpose in revealing to the world a new philosophy.

At most it is but one half the picture, and that the lower half. It exhibits the mouth only, the vehicle of the material things which sustain the body. Yet nevertheless not to be despised; for without it the body could not live, and without the body, the mind could have no connection with mortal minds, and as to these must be dead also. But it entirely cuts off and conceals the upper half of the man; the skull, the seat of mind, the residence of that God-inspired particle, which alone ennobles and makes valuable the whole body.

It is true that Bacon hoped by his philosophy to supply man's vulgar wants, and to make his sojourn here as easy and comfortable as was possible; but he sought this only as a necessary and blessed accident by the way, and not as the end of his new learning.

While he laboured to benefit mankind as mortal man, he also strove to elevate him as an immortal soul; mindful of the origin of which, he dared, like Plato, to hope to exalt man into a god, by leading the divine spirit, breathed into him when he was made in the image of God, to a contemplation and discovery of the secrets of the Great Artificer.

It was a favourite text of Bacon's, "It is the glory of God to conceal a matter; it is the glory of the King (a man) to find it out." (Prov. 25, 2).

Was not this very much like placing man almost on a parity with God, and exalting him into a god? And again, even misquoting to suit his lofty notions of man's capabilities: "The spirit of man is as the lamp of God, wherewith He searches every secret."* (Prov. 20, 27). Surely, too, the aim of him who describes the sole end of his philosophy in the following words, is not different from that of him who urges his disciples "to fix their minds on the contemplation of the immutable essences of things." "The end of our foundation is the knowledge of causes, and secret motions of

* Filum Labyrinthi.

things, and the enlarging of the bounds of human empire to the effecting of all things possible." *

Neither does he differ at all from the philosopher of the Academy in his appreciation of pure truth. " Truth, which only doth judge itself, teacheth that the inquiry of Truth, which is the love-making or wooing of it; the knowledge of Truth, which is the presence of it; and the belief of Truth, which is the enjoying of it; *is the sovereign good of human nature.* The poet saith excellently well: 'It is a pleasure to stand upon the shore, and to see ships tossed upon the sea; a pleasure to stand in the window of a castle, and to see a battle and the adventures thereof below; but no pleasure is comparable to the standing upon the vantage ground of Truth, and to see the errors and wanderings, and mists and tempests in the sea below;'† so

* New Atlantis.

† Bacon would seem to have had this passage again in his mind, when he described Plato as " a man of a sublime genius, *who took a view of everything as from a high rock.*" —*De Augmentis*, sec. 5.

always that this prospect be with pity, and not with swelling or pride. Certainly it is heaven upon earth to have a man's mind move in charity, rest in providence, and turn upon the poles of Truth."* Is this the language of one who had no higher aim than "to supply man's vulgar wants, and whose eye was ever on a mark which was placed on earth and within bowshot?" No! long since must Bacon have been forgotten, if his philosophy had had no higher end than that which modern utilitarianism deems its proudest boast.

One more extract will suffice to evince, that in promoting the proper study of his favourite science, Natural Philosophy, he had far higher views than mere utilitarianism; though this was to be regarded by the way and as an accident of no mean importance. "All knowledge, and especially that of natural philosophy, tendeth highly to the glory of God in His power, providence and benefits appearing and engraven in His works, which without this knowledge are

* Essay on Truth.

beheld but as through a veil, for if the heavens in the body of them do declare the glory of God to the eye, much more do they in the rule and decrees of them declare it to the understanding."*

An apology is needed for this long episode on Bacon, and our apology must be an anxious desire to direct the student back from the false school of Baconism to the master himself. Leave the Macaulays, the Herschels and the Playfairs to the work—and an important and useful work it is—for which they are fitted; but do you endeavour so to mind earthly things that you forget not heavenly things.† We say not, as did the ancient philosophers, disregard earthly things; but, while attending to them, forget not the heavenly, as the utilitarians do. Neither

* Filum Labyrinthi, Part I.

† Earthly and heavenly are not here used, in the New Testament sense, for sinful and holy; but in the Old Testament sense; earthly, for things pertaining to the body formed of the dust of the ground, and heavenly, for things pertaining to the mind, the breath of God.

would it have been necessary to have entered so fully into the matter had we not been aware that of the thousands who pretend to tread in the steps of Bacon, not above one or two have ever read his more important works; but take their notions of his philosophy from such crude and partial views as the merest utilitarians choose to enunciate as Baconian.

We require no other proof of the degeneracy of modern mind from the close habits of intense thought which distinguished the predecessors and cotemporaries of Bacon, than the melancholy fact, that while the *Novum Organum* and *De Augmentis* were, in the author's time, eagerly read by every one pretending to a liberal education, and at once elevated him to a high rank among literary men, they are scarcely ever opened in the present day, "and though much talked of are but little read. They have produced indeed a vast effect on the opinions of mankind, but they have produced it through the operation of intermediate agents."* Of these

* Macaulay's Essay on Bacon, vol. II., p. 426.

intermediate agents we have given a few specimens; and as long as the world submits to receive their version of Baconism, so long will Baconism elevate Knowledge at the expense of True Wisdom. Let men return to Bacon, and take *all* that he teaches instead of part—the inferior part—and there will be nothing for Wisdom or Knowledge to fear.

CHAPTER VI.

OF THE JEWISH NOSE.

CLASS IV.—THE JEWISH, or Hawk, Nose is very convex, and preserves its convexity, like a bow, throughout the whole length from the eyes to the tip. It is thin and sharp.

It indicates considerable Shrewdness in worldly matters; and deep insight into character, and facility of turning that insight to profitable account.

THIS is a good, useful, practical Nose, very able to carry its owner successfully through the world, that is as success is now-a-days measured, by weight of purse; nevertheless it will not elevate him to any very exalted pitch of intellectuality.

It is called the Jewish Nose in conformity with long-established nomenclature, and is, perhaps, more frequent among the Jews than among most other nations resident in Europe. It is, however, a fallacy to suppose that the peculiar physiognomy called Jewish is confined to the Jews, or even exclusively characteristic of them. It is in fact a form of profile common to all the inhabitants of Syria; and Sir G. Wilkinson has proved in his erudite work on Ancient Egypt, that the nations represented in the Egyptian sculptures with this cast of countenance are not always intended for Jews, as was at one time supposed, but for Syrians. Moreover, this form of countenance is to this day, the usual one among the Arabs of that part of the world. This Nose should therefore more properly be called the Syrian Nose.

This fact enables us to extend our illustrations, by adducing divers national proofs of the correctness of the indications ascribed to this Nose.

We have said that it is a good, useful, practical Nose, *i. e.* a good money-getting Nose, a good commercial Nose, and perhaps the latter term would be an apt secondary designation for it. Hence, those nations which have been most largely gifted with it, have been always celebrated for their commercial success.

The Phœnicians were Syrians, and the portraits which we have of these people on the Egyptian sculptures, as read by Sir G. Wilkinson, all exhibit this form of Nose. It is unnecessary to enlarge on the very early commercial activity of this nation, on its extensive traffic, its flourishing colonies, and its mighty fleets. While the rest of the world was in barbarism, or kept their low civilization carefully locked up within their own dominions, the Phœnicians were spreading arts and letters among the barbarous nations of Europe, and carrying civilization forward on its destined course towards the West. And the incentive to this and the means whereby it was effected were the same as those which now animate modern Tyre to promote the same Westward tendency of civilization. What Phœnicia, a little corner of Asia, did for Europe, England, a little corner of Europe, has done and is doing for lands still further West—America and Austral-Asia; destined to be in their turns the seats of a still progressive civilization, until every part of the earth shall have been in succession blessed with a civilization, if not always equal in degree, always adequate to its age, requirements, and capacity.

Then when the whole circle shall have been accomplished—and of which more than two-thirds have been already passed over—when civilization in Austral-Asia shall touch the confines of its original starting-point, the Eastern shores of India, the consummation of all things shall be at hand; the purpose for which the earth was created, and for which millions of years have been slowly, surely, and silently beautifying, storing, and adapting it, until it is like "the Garden of the Lord," shall have been fulfilled; and the whole of this beautiful system shall vanish away like a breath, yet leave no vacuity, no defect, in the vast and mighty universe, whose limits utterly transcend our notions of time and space.

Two-thirds of this circle have been already

passed over; the remaining third is rapidly running out; we already stand half-way between the beginning and the end of this third part; nay, we are nearer the end than the beginning; we see more clearly and apprehend more closely the day when Austral-Asia shall be the seat of civilization and Christianity, than we do the day when those blessings seventeen hundred years ago, first landed on our shores; we feel more affinity for, and more sympathy with the latter age than with the former, and we may be assured that we do this because we are much nearer in Time to the one than to the other.

This is an awful contemplation; we cannot but feel that there is an extra responsibility cast upon us upon whom literally "the ends of the world are come," and that it concerns us more than all who have gone before to be up and be doing; to take heed that while civilization is progressing geographically, it is also progressing in power and character; for upon the extent and nature of the Knowledge which we transmit, depend in a great degree the extent and nature of the Knowledge which shall ever reign on the earth.

Theologically considered, the subject is infinitely more awful and important; and the mind cannot contemplate without fear and trembling, what may be the consequences if we, instead of a pure and perfect, transmit to the few generations yet to subsist on the earth, an impure and imperfect Christianity.

But to return to our more immediate subject. The Jews have always been celebrated for shrewdness in commercial affairs. Though the peculiarities of their religion prevented them from taking a leading part in the general commercial business of the ancient world, yet among themselves trade always flourished; and in the present age of the world, the Jews were in all countries the first revivers of commerce after the stagnation occasioned by the irruptions of the northern hordes, and in many nations are still almost the only traders.

It does not always follow, however, that the love and capacity for getting money is accompanied by a sordid disinclination to part with it.

Numerous instances occur of persons who shrewdly bargain for pence, but liberally give away pounds. As we may seem to have inferred that the former is a Jewish habit, it is right, and we are happy to be able to say, that some instances of princely liberality among modern Jews, afford lessons which Christians would do well to take.

No very exalted intellectuality is to be looked for from the Syrian nose. Its sphere of action is widely different from that of mental exertion for the mere pleasure thence derivable. Hence, we find, that notwithstanding the free intercourse which the Phœnicians permitted with all nations, the ancient sages rarely travelled to Phœnicia for learning. If they went there, they went like Solomon, to traffic. They sought learning among the Chaldeans, the Indians and the Egyptians, but seldom touched in their course on the more accessible shores of Phœnicia. The Phœnicians have had the reputation of being the inventors of letters because they introduced them into Europe; but they were the mere carriers of them for commercial purposes, not the inventors.

Though some attempts have been lately made to prove that the Hebrew nation has furnished more learned men than any other, the attempts are an utter failure.

Curious wranglers, ingenious cabalists, fine splitters of hairs, shrewd perverters of texts, sharp detecters of discrepancies, clever concocters of analogies, finders of mysteries in a sun-beam, constitute the mass of modern* Jewish scholars. What is the Talmud, the Mishna, the Gemara, or any of their comments thereon, or on Scripture, but mere puerile exercises of wit; sometimes ingenious, but always reckless of truth, decency or common sense? We search in vain, as far as our knowledge of their works extends, and as all those who have studied them assert, for any expanded views, any comprehensive ideas or extensive learning. Neither does their ancient history furnish any but inspired names, to class among the world's sages.

Education is however rapidly extending among

* *i. e.* Post Christum.

the Jews. For the first time since they ceased to be a nation they appear to begin to feel the importance of raising themselves to an equal intellectual rank with the citizens among whom their lot is cast.

Numerous schools have recently been founded by them for the education of their own people —both male and female — in England and other European States. From these the most beneficial results may be anticipated.

It has always been found to be the greatest obstacle to the spread of Christianity among a people who *a priori* might be supposed to be the most ready to receive it as a proof of the truth and fulfilment of their own Scriptures, that they know not these Scriptures; but are either immersed in the grossest ignorance, or glean their religion from the Talmud and the Mishna. It has been justly said, "The Jews must be made Old Testament Jews before they can be made Christians;" and this can only be done by education among themselves creating a spontaneous spirit of inquiry into their own literature, with an anxious desire to read and comprehend the vast storehouse of Biblical treasure at present almost unknown to the large majority of them.

The sources of our individual illustrations treating only of those who have distinguished themselves in Literature or History furnish only a few examples of the Jewish Nose.

Vespasian,
Correggio,
Adam Smith,

may serve, however, to illustrate and corroborate our theory. As to the last, the connection

ADAM SMITH.

between his Nose and the peculiar bias of his mind is obvious.

"The founder of the Science of Political Economy" must have possessed a natural attraction towards commercial affairs; and it could only have been by a very large share of acute observation and shrewd penetration that he could have worked out the principles of so abstruse a science, and made it acceptable to the mass of mankind.

"It was," says one of his admirers, "one of the few, but greatest, errors of Adam Smith, that he was too apt to consider man as a mere *money-making* animal, who will never hesitate to work provided he is well paid for it. He does not consider that the desire of power and of esteem are more powerful principles than the desire of wealth."

It is impossible to desire a description of his character more exactly correspondent to the form of his Nose.

It has been much disputed among his biographers whether CORREGGIO was rich or poor. Many anecdotes are related which indicate his extreme poverty; while on the other hand, numerous facts seem to prove that he must at least have been in easy circumstances.

He married a lady of good fortune, and he was well appreciated in his own time, and received many valuable orders for paintings from patrons of high rank and great liberality. It is however undisputed that his disposition was penurious and miserly, and this fact—indicated also by his unusually well-developed hawk-nose—will serve to reconcile the apparently contradictory assertions of his biographers.

CORREGGIO.

It is probable that, like most misers, he was always complaining of poverty, and even denied himself necessaries which he could have well afforded. Those who credited these complaints, recorded his poverty and lamented over it with mistaken kindness; while others, who more critically considered his actual means, would better appreciate them and reveal the true state of the case. There is an anecdote recorded of him by his friend and cotemporary, Vasari, which though it may not be wholly true, has probably some foundation. It is not, however, as Gibbon has shrewdly remarked, of much importance whether an anecdote of a person is actually true or false; for it almost equally displays the character of the person of whom it is recorded. A tale of liberality is not told of a known miser; nor an instance of penuriousness of a liberal man. An anecdote, to be received, must at least be probable and have an air of verisimilitude. Neither, considering the character of Correggio, is there any such inconsistency in the story as to render it incredible. The objection that sixty crowns in copper would weigh two hundred pounds, and therefore be an impossible weight for a man to carry, is a mere quibble. It only proves that the quantity is exaggerated, and not that the main story is false.

This characteristic anecdote is to the effect, that having received a payment of sixty crowns in copper, he carried it home on foot in sultry weather, and the over-fatigue brought on a fever, of which he died.

VESPASIAN.

*(From a coin in the Museum of Florence.)**

The character of VESPASIAN has been painted in the brightest colours. Avarice alone sullied

* This head enables us to point out a characteristic difference between the convexity of the Jewish Nose and the Roman. The convexity of the former commences at the eyes, and if afterwards it aquilines, the Nose is $\frac{\mathrm{I}}{\mathrm{IV}}$ or $\frac{\mathrm{IV}}{\mathrm{I}}$, according as I. or IV. prevails. The convexity of the Roman Nose is confined to the *centre* of the Nose, and occasions its aquilineness.

his virtues. This must have been no slight or temporary blot, or his eulogist and client, Tacitus, would not have recorded it. It was too palpable and notorious to be concealed, and the historian found himself, however reluctantly, compelled to confess it.

It is not improbable, that he inherited this vice; for his father, having saved money in the business of a collector of the revenue and retired from the office, was unable to resist the love of gain, and subsequently acquired a considerable fortune by lending money at usurious interest. The prudence and sagacity with which the young Vespasian regulated his conduct during the dangerous reigns of the brutal Caligula and Nero, indicates his penetration and sagacity. It must have been by no trifling tact and ingenuity that he escaped death for the heinous offence of appearing inattentive while the Emperor Nero was singing. The same shrewdness and insight into character enabled him while in a private station to redeem his ruined fortune by horse-dealing; a science always notorious for its unscrupulous scheming and dishonest sharp practice; and in which the hawk-nosed Syrian Arabs have ever excelled all other nations.

TITUS, the successor and son of Vespasian, inherited his father's profile, and it is a marked corroboration of our theory that avarice is the only vice attributed to that otherwise virtuous prince.

It must however be observed, that the Noses, both of Vespasian and his son, were not purely Jewish, but *Judæo-Roman* $\frac{IV}{I}$; a formation which corresponds accurately with other peculiarities in the characters of those great generals, too well known to need further elucidation.

CHAPTER VII.

OF THE SNUB NOSE AND THE CELESTIAL NOSE.

CLASSES V. and VI.—THE SNUB NOSE AND THE TURN-UP (*poeticè*) CELESTIAL NOSE.

The form of the former is sufficiently indicated by its name. The latter is distinguished by its presenting a continuous concavity from the eyes to the tip. It is converse in shape to the Jewish Nose. N.B. It must not be confounded with a Nose which, belonging to one of the other Classes in the upper part, terminates in a slight distension of the tip; for this, so far from prejudicing the character, rather adds to it warmth and activity.

We associate the Snub and the Celestial in nearly the same category, as they both indicate natural weakness; mean, disagreeable petty disposition; with petty insolence, and divers other characteristics of conscious weakness, which strongly assimilate them (indeed, a true Celestial Nose is only a Snub turned up); while their general poverty of distinctive character, makes it almost impossible to distinguish their psychology. Nevertheless, there is a difference between their indications; arising, however, rather from degree than character. The Celestial is, by virtue of its greater length, decidedly preferable to the Snub, as it has all the above unfortunate propensities in a much less degree, and is not without some share of small shrewdness and fox-like common sense; on which, however, it is apt to presume, and is, therefore, a more impudent Nose than the Snub.

It is with considerable distaste and rèluctance that we approach the latter divisions of our Classification. *Pœnitet me hujus Nàsi.* We wish we had never undertaken to write of these Noses. Having done so, however, we must fulfil our engagement. But the mind shrinks from the thought, that after contemplating the powerful Roman-nosed movers of the world's destinies, or the refined and elegant Greek-nosed arbiters of art, or the deep and serious-minded thinkers with Cogitative Noses, it must descend to the horrid bathos, the imbecile inanity of the Snub.

Perhaps the reader expects that we are going to be very funny on the subject of these Noses. But we are not;—far from it. A Snub Nose is to us a subject of most melancholy contemplation. We behold in it a proof of the degeneracy of the human race. We feel that such was not the shape of Adam's Nose; that the original type has been departed from; that the depravity of man's heart has extended itself to his features, and that, to parody Cowper's line, purloined, by the bye, from Cowley:—

"God made the *Roman*, and man made the *Snub*."

Fortunately for our hypothesis, and for our feelings, we cannot find a single instance of the existence of either the Celestial or the Snub among celebrated persons, except in those who are illustrious by courtesy rather than by their actions, and whom station, not worth, has made conspicuous. The following are the only instances of the Celestial Nose which our pictorial sources furnish:—

James I.
Richard Cromwell.

Mary, wife of William III.
George I.
Kosciusko.

KOSCIUSKO.

Peculiar circumstances won Kosciusko somewhat of a name, for it was rather from sympathy with his cause than from admiration of his abilities, that it was ever bruited in men's mouths, or is yet remembered. Had he been gifted with a Roman Nose, that is, had his soul been Roman, energetic, dignified and self-reliant, Poland might have risen again into the rank of nations. But he submitted to crouch beneath the rod of Napoleon, temporizing and treating for benefits for which it was his duty to have fought; and the nation, which looked to him for assistance, was compelled to share his degraded fate, and become the despised tool of an all-grasping despot. He had, however, a share of the Cogitative with the Celestial; and thus affords an instance of an union so rare, that it is only to be regarded as an exception to the rule laid down, that Class III is never associated with V and VI.

From fictitious works, which have raised to celebrity imaginary characters of every mental calibre, innumerable examples might be adduced; for all accurate observers, whether ancient or modern, have—without being professed Nasologists—unconsciously verified our hypothesis, and associated the Nose with character.

The inimitable Dickens, and his equally clever illustrator Cruikshank, both of whom owe their power to their correct observation and delineation of character, afford many well-known examples. Had the hypothesis been founded on Oliver Twist and its illustrations, it could not have been more strikingly substantiated by them, than it is—thus proving that if we err, we err in company with observers of more than common accuracy, and whose observations have been verified by the applauses of all. In that work we have the shrewd penetrative Jew with his Hawk-nose; the mild, but high-minded Oliver Twist, with his fine Greek nose; the Artful Dodger and his brother-pals with their characteristic Snubs and Celestials. A reference to the plates, and the author's pen-and-ink portraits, in this and other works, will confirm our right to claim Dickens as a Nasologist.

The only authority which we have consulted on the subject of Noses, is one from whose works we have already quoted. It never can be forgotten that the inimitable Tristram Shandy has slightly touched upon the subject when describing the unhappy catastrophe which, even in his very earliest years, demolished his Nose.

It appears that Mr. Shandy senior, was a sagacious, an observant, and a learned man. We need not add, therefore, that he was deeply imbued with the importance of his son having a good Nose; and most pathetic was his sorrow when the bridge of it was broken. His own family had suffered through several generations from a defect in the length of an ancestor's Nose. His great-grandfather, when tendering his hand and heart to the lady who afterwards consented to make him "the happiest of men," was forced to capitulate to her terms, owing to the brevity of his Nose.

"It is most unconscionable, Madam," said he, "that you, who have only two thousand pounds to your fortune, should demand from me an allowance of £300 a year."

"Because you have no Nose, Sir."

"'Sdeath! Madam, 'tis a very good Nose."

"'Tis for all the world like an ace-of-clubs."

"My great grandfather was silenced:" and for many years after the Shandy family was burdened with the payment of this large annuity out of a small estate, because his great great-grandfather had a Snub Nose. Well might Mr. Shandy (the father of Tristram) say "that no family, however high, could stand against a succession of short Noses!"

In lack of other instances, we have introduced those of fictitious writers; for they corroborate our views, and serve to thicken other proofs which in this Class do demonstrate thinly. And this necessarily so. For we have determined to refrain from giving examples from our personal acquaintance, and the Snubs have never any of them won such eminence, as to have their names handed down by fame, or their portraits limned for the benefit of posterity. The evidence in these two last Classes is necessarily negative.

Their best proof lies in their want of proofs. The Snub will, however, receive some general illustration when we come to speak of national Noses.

It now only remains to treat of some obstinate Noses which will not come within our classification.

One of these is that curious formation, a compound of Roman, Greek, Cogitative, and Celestial, with the addition of a button at the end, prefixed to the front of my Lord Brougham. We are bound from its situation to admit that it is a Nose, and we must, therefore, treat of it; but it's a queer one. "Sure such a Nose was never seen."

It is a most eccentric nose; it comes within no possible category; it is like no other man's; it has good points, and bad points, and no point at all. When you think it is going right on for a Roman, it suddenly becomes a Greek; when you have written it down Cogitative, it becomes as sharp as a knife. At first view it seems a Celestial; but Celestial it is not; its Celestiality is not heavenward, but right out into illimitable space, pointing—we know not where. It is a regular Proteus; when you have caught it in one shape, it instantly becomes another. Turn it, and twist it, and view it how, when, or where you will, it is never to be seen twice in the same shape, and all you can say of it is, that it's a queer one. And such exactly is my Lord Brougham; —verily my Lord Brougham, and my Lord Brougham's Nose have not their likeness in heaven or earth—and the button at the end is the cause of it all.

Thus, though Lord Brougham's Nose is an exception to our classification, it is not, as has been asserted, an exception to our system. On the contrary, it is manifestly a strong corroboration of it. The only exceptions are those where the *character does not correspond with the Nose*, and of those we have yet to hear.

There is another Nose which is not included in the classification, but which, though not peculiar to *one* individual, is nevertheless not sufficiently frequent to demand placing there. This we call the Parabolic Nose. It would have been a good Nose if it had gone on as it began; but, having from some cause taken an inward curve too soon, its good qualities become nearly nullified. It presents a continued Parabolic curve, where it ought to extend into an angular tip. This sudden abbreviation of course weakens the character, but, as it leaves the good qualities of the upper part still inherent, the character retains good points; but being disabled from reasoning justly on its good intentions, it acquires the character of obstinacy, and of acting from pig-headedness, instead of from rational forethought.

GEORGE III. presents the best-known example of this Nose.

Another striking example occurs in BLANCO

WHITE. There were considerable points of identity between their characters.

They were both honest, conscientious men, anxious to find out and pursue the right course, but both were too hasty in jumping to conclusions to form accurate judgments. Blanco White, anxious to embrace truth, led a regular harlequin dance after her all his life, and died in motley. One leg red and the other blue, with a jacket of various colours, and a coxcomb of brilliant self-conceit. His last verdict, after rambling through divers forms of religion and no-religion, was, "I am neither Trinitarian, nor Unitarian, nor yet Arian." First Roman Catholic, then Atheist, then Church of England, then Unitarian, then Arian, then Omniarian, his ardent, hasty mind settled like a butterfly on the first bright flower which fluttered in the breeze, for a time imbibed and luxuriated on its honey, and then flew off to suck the sweets of some other plant. Thus he fluttered on, a varied, anxious, unsettled existence, gathering honey, but making none; and when the colds and storms of winter came, he sank before them.

The instances of the Parabolic Nose are, however, too few to justify deductions from it, and we would rather, at present, not express decidedly what are its indications. Should we be able to do so at any future time it will be entitled to stand as Class VII.

CHAPTER VIII.

OF FEMININE NOSES.

The subject of Nasology would not be complete without some observations on the Feminine Nose, because sex modifies the indications, some of which, though disagreeable and repulsive in a man, are rather pleasing, fascinating and bewitching in a woman, and *vice versâ*.

It is the fashion for women to aspire to equality with the other sex, and as long as they will be content with an equality, in a different orbit, they are undoubtedly entitled to it. It should, however, be the equality of planets—each perfect and beautiful, each useful and beneficial in its sphere; but pregnant with disorder and confusion when Venus would invade the orbit of Jupiter, or intrude within the circuit of Mars.

No intelligent man denies to woman such an equality; but as certainly as a good housewife would pin a dish-cloth to the coat-tail of a husband prying into the mysteries of the kitchen and claiming equality with his wife in the household sphere, so surely will men cry out against and turn with disgust from women who invade their province of warriors, statesmen, merchants, &c.

Nevertheless, let us not be misunderstood, or be accused of including in a sweeping clause those cases which are, of right, exceptions. A woman may be placed in such a position that active life is her legitimate sphere, and that if she neglects or devolves its cares upon others she is culpable. We all feel an enthusiastic respect for the noble Boadicea, arousing her pusillanimous countrymen against the cruel ravages of the Romans, and dwell with admiration on Elizabeth haranguing her army at Tilbury and personally engaging in affairs of State, because they were occupied in duties which became a monarch; yet if a woman, who has no call to any higher duties than those of domestic life, were to leave them to engage in the contests of warriors or the turmoil of politics, we should regard her as an unfeminine virago. Notwithstanding, though the woman may in some cases be needfully sunk in the station, those duties which become the former will still engage more of our love and regard than those which belong to the latter; and our own graceful Queen has secured, by her happy union of the duties of both, more of the love and respect of her people than any of her predecessors on the throne of these realms.

The energies and tastes of women are generally less intense than those of men; hence their characters appear less developed and exhibit greater uniformity. That their passions are stronger is undeniable, but these do not constitute character, nor are exhibited in the Nose. Their indexes are the eyes and mouth, and therefore their consideration forms no part of the present subject. This uniformity of character is noticed by Pope in a line which at first sight reads libellous, either because it appears to refer to moral conduct—which it does not—or because it is too sweeping and exaggerated. He asserts roundly,

> "Most women have no characters at all."

No characters at all is obviously false; but, as compared to men, as near the truth as most general epigrammatic rules are. It is in the latter sense that Pope used it to illustrate the difficulty of discussing "The characteristics of Women" after a dissertation on those of men. The line, however, was truer in his time than it is now, when more general and more liberal education has tended very much to break up the uniformity of character which existed among the inane ladies of Pope's era.

Nevertheless, whether repressed by Art or curtailed by Nature, women's characters certainly *appear* less developed than those of men. If by Nature, it is a blessed provision—as all nature's providings are. It is the woman's place to be in rational subjection to the man; and though the sweet saints would sooner tear out the eyes of St. Paul* (we wonder he is such a favourite with them) than confess his precepts in terms, yet they do not fear to acknowledge that they have no respect for the man who succumbs to his wife, or admiration for the woman who aspires to denude her husband of his appropriate symbols of masterdom.

If this happy inferiority—an inferiority which places them far above men in practical wisdom, inasmuch as it consists in shrewd, practical common sense, against man's intellectual blundering—if this happy inferiority is the result of Art, they exhibit in its adoption much sound wisdom. Man is an insolent, domineering, self-sufficient animal—let him *say* what he will about the elevation of the female mind, we believe no man ever fell in love with the woman whom he felt to be wiser than himself. He could not endure for a partner for life, such a perpetual looking-glass, and reminder of his own infirmities; he could not bear the constant attestation of his own weakness. He

* Enhes. v.. 22—24.

could regard patiently the vaunted accomplishments of another man, but he could not submit that his wife should be his acknowledged superior, and to be her foil—perhaps fool.

Hence it is that wise men so frequently, that it is become proverbial, marry silly women. However much a learned man may admire female accomplishments, he detests a woman who strives to rival him in his own sphere, who is talking philosophy when he would be whispering "soft nothings," and who freezes his ardent admiration with a dissertation on mathematics, or a moral discourse on self-control. He can bend, like any other man, with intense joy, over the blushing girl who tremblingly believes that her eyes are brighter and more lovely than the stars over her head; but would fling from him with disgust the woman who would repress his harmless and true—because soul-felt—flattery, with a philosophical disquisition on the nature, distances, and offices of the aforesaid stars. And it is because learned women too often strive by this injudicious ill-timed wisdom, to catch learned men for husbands, (and there are no more determined husband-hunters than blue-stocking women, because they are always within a year or two of being shelved), that the latter are necessarily flung into the arms of women who they know *can't* bore them with an eternal round of sense, from which every one is glad occasionally to escape, and never more so than when he is in love.

Hence it is that blue-stocking women are proverbially avoided by men; not because men despise or dislike their learning, but because they make such ill-timed use of it. They may be admired, but they are never loved; they may talk as wise and as learned as is in their power, but learning and wisdom never won a lover, much less a husband. *Ver. sap.* my dear lady reader, and if you don't understand the abbreviate, ask—ask—anybody, but your husband.

> "Yes Love, indeed, is light from heaven,
> A spark of that immortal fire,
> By angels shared, to mortals given,
> To lift from earth our low desire."

And shall heaven-born love bow to mortal wisdom? Shall the God whom Jove himself obeys, become the slave of Minerva? No! let Love wear the cap and bells of Folly, but shroud him not in the cold cerements of the Goddess of Wisdom! Be assured, the doves of Venus will never nestle under the dusky wings of the sage owl of *innupta* Minerva, who, herself, could never win a husband, or a lover, from the whole host of Olympus.

Whatever the cause, it is almost indisputable that women's characters are generally less developed than those of men; and this fact accurately accords with the usual development of their Noses. But for a small *hiatus* in the prosody, Pope's line would read equally well thus:—

"Most women have no Noses at all."

Not, of course, that the nasal appendage is wanting, any more than Pope intended by the original line that women's characteristics were wholly negative; but that, like their characters, their Noses are, for the most part, cast in a smaller and less developed mould than the Nose masculine.

In judging of the Nose feminine, therefore, comparison must not be made with the masculine, but with other feminine Noses. All the rules and classifications apply to the one as well as the other, but allowance is to be made for *sex*.

The Roman Nose largely developed in a woman mars beauty, and imparts a hardness and masculine energy to the face which is unpleasing, because opposed to our ideas of woman's softness and gentle temperament. In a man we admire stern energy and bold independence, and can even forgive, for their sakes, somewhat of coarseness; but in a woman the former are, at the least, unprepossessing and unfeminine, and the latter is utterly intolerable. Woman's best sustainer is a pure mind; man's a bold heart.

Moreover, the exhibition of character in women should be different from that in men. From the masculine Roman Nose we may justly look for energy in the active departments of life, but in a woman its indications are appropriately exhibited in firmness and regularity in those duties which legitimately fall to her lot. We do not desire to see a woman so endowed launch out, uncalled for, into the bustle and turmoil of the world, or endeavour to take the reins of government from her husband, though she may be equally well fitted for the task: but we are content to see her govern her household with energy, and train up her children in a systematic and uniform manner.

She will form her plans of household management with promptitude, and carry them out with undeviating firmness and decision: and her husband will act wisely, for his own sake, not to interfere with her, so long as her energy does not carry her into his department.

But if woman's circumstances place her in a more extended sphere, her career will afford an example to illustrate our hypothesis as well as that of a man. Of this we have an example in the illustrious Roman Lady, Livia, the wife of Augustus.

LIVIA.

(From a coin in the Museum of Florence.)

Her nose presents a combination of the Roman and the Greek, and contains as much of the former class as is compatible with female beauty. The accounts which are handed down concerning her are very contradictory: some describing her as chaste as the icicle that hangs on Dian's temple, and qualified to lead a chorus of vestals, while others accuse her of licentiousness and criminal amours. It is, however, undeniable that she was a woman of considerable power of mind, which she exercised energetically and shrewdly in procuring the aggrandizement of her son Tiberius, on whose head she finally succeeded in placing the imperial tiara. Her Roman energy was nevertheless refined by an infusion of Greek elegance, and she was a liberal patroness of arts and literature. Her career likewise illustrates another maxim; that what woman's character wants in development, is often compensated by superior passion. Livia was sustained more by the strength of her affections than by personal ambition. It was her son's and not her own aggrandizement that she sedulously pursued; and if the lives of the majority of ambitious women were examined, it would be found that they more frequently sought to exalt some object of their affections—a husband or a child—than themselves.

This, however, was not the case with the purely Roman-Nosed Elizabeth. She had no affection for any one but herself; and the energy and determination, combined with the coarseness of her character, correspond accurately with the indications of her Nose.

The most beautiful form of Nose in woman is the Greek. It is essentially a feminine Nose, and it is in its higher indications that women generally excel.

This Nose will not carry them out of their natural sphere, and it is for this reason that it is so beautiful. Congruity is harmony; and harmony is essential to the beautiful. A woman gifted with the feelings of a poet, need not fear to give them full sway. In some of the most beautiful and touching departments of poetic talent women equal—perhaps excel—men. Scarcely half a century has elapsed since women were permitted to cultivate unreservedly the fields of literature, but that brief period has incontrovertibly proved the ability of women to pourtray with superior truth and pathos all that relates to the affections, the sentiments, and the moral and religious duties of mankind.

The names of Hannah More, Barbauld, Edgeworth, Tighe, Hemans, De Stael, and other lamented writers, together with those of several who still survive, place this assertion beyond the pale of controversy. The Noses of the above-named gifted women were Greco-Cogitative.

MRS. HEMANS.

But the power of expression, though essential to a poet, is not necessary to a poetic mind. It may exist as strongly in one who has no words of fire to give its creations utterance as in one who pours forth in lavish self-abandonment the riches of his soul. Neither is the Greek Nose a necessary index of a poetic faculty. That form may adorn the face, but no rapturous fervour exalt the mind; although it will frequently accompany a poetic temperament, because it indicates refinement and purity of taste. These are its invariable indications, and in these every woman so gifted will excel; for to excel in these is almost her peculiar province.

In the minor and domestic departments of life, where woman's influence is so peculiarly blessed, the refinement of the Greek Nose will appear in those household arrangements which make home the happiest and most beloved spot on earth. It will exhibit itself in her needle-work by an artistic arrangement of colours and a poetic choice of subjects; in a neat and elegant attire, in the decoration of her drawing-room, or in the paraphernalia of her boudoir. Nor need it be confined to those elegancies which seem to belong exclusively to the higher classes—a cup of flowers in a cottage window, the well-selected trimmings of a Sunday cap, or a pretty ornament on the mantel-shelf will equally be an evidence of a refined taste, and be found to accompany a Greek Nose.

The Cogitative Nose does not so frequently appear among women as among men. Women rather *feel* than think. Their perceptions are intuitive, instinctive; men's Cogitative. They are shrewder and more instantaneous in estimating character, or in deciding on action than men. Men must think, and fume, and fret before they can decide; must, in common parlance, set the head (reason) against the heart (instinct); while women rely more on the latter, and are consequently, in judging of character or in deciding on a course of moral conduct, more frequently right than men.

Our advice to a man would be this: if you are at a loss, after long cogitation,—as ten to one you will be—to know whether an intended act is morally right, ask a sensible woman, and she will guide you with perfect wisdom in a minute. So again: if you would know any one's moral character, let a sensible woman converse with him for five minutes and she will tell you without fail whether he may be trusted. Only be careful and accept her first dictum; don't argue the point with her, nor give her time to *think*; have her instinctive decision. If she thinks, she will be ten times more at fault than a man; and, if you argue the matter with her, she will lead you a dance through as fine a quagmire of absurdities as can be conceived, and there leave you, up to your neck in the slough, without the power—if not without the will—to help you out. And this needfully so. Instinct must ever be a better guide than Reason: for,

> "In this (Instinct) 'tis God that acts, in that (Reason) 'tis man."

"The perception of a woman," says Sherlock, "is as quick as lightning. Her penetration is intuition, almost instinct. By a glance she will draw a quick and just conclusion. Ask her how she formed it and she cannot answer the question. While she trusts her instinct she is scarcely ever deceived, but she is generally lost when she begins to reason." A more accurate picture of the female mind was never drawn; yet some modern writers have fiercely controverted it. Under a mistaken notion of equalizing women with men, they seek to destroy the individualism of their character. One witty popular writer has even ventured to assert, that if half a dozen boys were brought up as girls, and half a dozen girls as boys, the latter would be to all intents psychologically men, and the former psychologically women. Surely a more preposterous absurdity never won the assent of the unthinking part of the community; nevertheless, it has been warmly applauded and often repeated, as if it were an ascertained fact instead of a ridiculous fancy.

The Jewish Nose is not very frequent among women. Neither are its indications material to the perfection of the female character. It is the duty of men to relieve women from the cares of commercial life, and to stand between them and those who would impose upon their credulity. Moreover, woman's natural penetration supplies the want of the thoughtful sagacity which protects men in intercommercial relations.

The remarks which we made on the Snub Nose and the Celestial Nose in men require to be considerably modified when we treat of those classes in women.

We confess a lurking *penchant*, a sort of sneaking affection which we cannot resist, for the latter of these in a woman It does not command our admiration and respect like the Greek, to which we could bow down as to a goddess, but it makes sad work with our affections. The former too is not so unbearable as in a man. It is a great marrer of beauty undoubtedly; but merely regarded as an index of weakness it claims our kindly consideration. Weakness in a man is detestable, in a woman excusable and rather loveable. It is a woman's place to be supported, not to support. Hence the classical emblem of the Vine and the Elm is felt to be beautiful and true, because it pourtrays accurately the natural mutual position of husband and wife. A woman, moreover, has generally tact sufficient to conceal (often to their entire annihilation) those unprepossessing characteristics of the Snub and the Celestial, which in a weak man become every day more and more strongly marked. A woman's weakness too is rather flattering, as it attests our supremacy; a thing which we like to be constantly reminded of, and of which we are very jealous, as it stands on rather ticklish and much disputed ground.

The impudence too, which is utterly unendurable in a male Celestial, and which seems to court contact with the toe of one's boot, is in a woman rather piquant and interesting. A Celestial Nose in a woman is very frequently an ,index of wit. Wit is a talent not emanating from wisdom; quite the reverse. The wisest men are ofttimes the slowest. Wisdom comes after thought, wit before it. A Celestial-nosed woman is only more witty than a similarly gifted man, because the impudence which it invariably indicates is backed by woman's ever-ready tact and quickness.

The indications are not varied; but the exhibitions are. Even if a man were gifted with the power of uttering the severe witticisms, and cutting repartees which are nectar and ambrosia from the lips of a pretty woman, he dare not; for he would be inevitably kicked down stairs—if the fellow were worth the exertion.

In a witty woman who can skirmish with unflinching quickness and dexterity, we can even forgive a slight moral delinquency. A little white-lie simpered out with arch assurance by a pair of demure lips,

"Like leaves of crimson tulips met,"

by no means offends us as it would in a man; in whom we should attribute it to low cunning or mean cowardice. Indeed the exquisite look of arch impudence with which a delicately chiselled marbleine Celestial tells you a most palpable falsehood is maddening, perfectly beautiful, almost sublime. The cool assurance and sharp raillery with which she persists after detection! the assumption of injured innocence! the impudent look of defiance! By Jove! truly

> "The dear creatures lie with such a grace,
> There's nothing so becoming to the face."

And then when they are beaten from their last defence, and can resist no longer, when they are compelled to surrender and beg pardon, they do it as if *they* were forgiving *you*; and make you feel almost as if you were being forgiven, as if you, not she, had all the while been erring: at all events you feel very like a fool, though very happy; and so a few tears, and a few (or not a *few*) kisses set all to rights,

"And so we make it up;
And then—and then—and then—sit down and sup."

All things considered therefore, and inasmuch as we prefer the naturalness of a witty woman to the artificialness of a learned woman, we confess to a liking for the Celestial Nose feminine, while we abhor the masculine. It is not, however, every female Celestial Nose that we admire (Heaven for our peace's sake forbid —they are so numerous). It must be of the purest and most delicate chiselling; have no tendency to cogitativeness, lest it should look as if its owner thought; and its hue must be of the palest and most evanescent flesh-tint. These are essential to indicate that delicacy of mind which alone makes wit in a woman fascinating, and which pardons breaches of strict morality committed from the purest and most benevolent intentions.

This sounds rather paradoxical, but an old Jacobite song will illustrate our meaning. The story goes that a gude-wife concealed a north country cousin, one of the adherents of Charlie, in the house unknown to the gude-man; and her ingenuity is sorely puzzled to account for certain suspicious phenomena which strike him on his coming home:—

"Hame came our gudeman at e'en,
  And hame came he,
And there he saw a pair o' boots,
  Where nae the boots should be.

'And how came these boots here,
  And whase can they be?
And how came thae boots here
  Without the leave of me?'
    'Boots!' quo' she; (*with amazement*)
    'Aye, boots!' quo' he.

'Ye auld blind dotard carle,
  And blinder mat ye be! (*indignantly*)
It's but a pair o' water-stoups,
  My minnie sent to me.'
    'Water-stoups?' quo' he,
    'Aye, water-stoups;' quo' she."
      (*with impudent determination*).

And so in like manner she unblushingly persists, in order to preserve her guest's life, that a saddle-horse is a milking cow, and a man's coat a pair of blankets. Now we are sure this dear woman had a Celestial Nose; nothing else would have had the ready wit and the impudent assurance to attempt to befool her gudeman, and to persist, with the addition of no slight abuse of his dotard blindness, in her palpable falsehoods; yet we defy any one not to love the good woman, and excuse her breaches of morality for the sake of her hospitable benevolence.

We are conscious that in discussing female Noses, we are treading on delicate ground. It is a difficult and nervous subject. We have endeavoured, however, to say nothing but what appeared to us to be plain truth. Nevertheless we would apologize if we have given offence to any one, were it not that we forcibly feel the truth of the homely adage, "the least said the soonest mended," and therefore hasten to close a chapter which has given us more trouble and anxiety than all the rest together.

CHAPTER IX.

OF NATIONAL NOSES.

THE reader will probably have been led from the nomenclature, to inquire whether the assertion that certain forms of Nose are justly named after certain nations might not be extended further? and whether every nation has not a characteristic Nose?

The reply to these questionings would be in the affirmative. Every nation has a characteristic Nose; and the less advanced the nation is in civilization, the more general and perceptible is the characteristic form. While nations are in their infancy, and the mass of the people are uninformed, the features, receiving no impressions from within, take the form impressed from without, and follow the national type. If one uniform state of things—of government, climate, and habits—continue, without education, generation may succeed generation and the original facial type of the race will remain. If, however, the national circumstances alter (still without general education) the national features follow the type impressed by those circumstances. We have appealed to many instances of these simultataneous national changes when describing the different forms of Noses prevalent at different periods of English history.

When however education becomes general, nations lose these national typical features; for the physiognomy becomes so variously impressed from within, according to the different bias and affections of men's minds, that it ceases to receive those impressions from without, which generate national types. At present, however, there is so little generally diffused education that the typical features of most nations may yet be defined.

These are not always the original types of the race. Numerous circumstances have among the more civilized nations contributed to produce changes of greater or less magnitude. The various Caucasian nations, for instance, though all descended from one stock, have mostly lost their original type in their various migrations from the plains of Asia, and received such typical form as varying circumstances have since impressed. Hence the various Caucasian nations of Europe and Western Asia differ considerably from each in mental and bodily organization.

These variations from the original type took place, however, at so early a period, even in the ante-historical period, that historians are apt to regard them as original and innate; and perhaps it is most convenient for *them* to do so. But this is not sufficient for the inquirer into the Races of Men. He goes back to ages far beyond the historical, or even the mythic, period; and, finding these nations are descended from one family, perceives that the present variations must have taken place after the dispersion of the family into distant localities under leaders of very various temperament and views of social happiness.

It would lead us too far to inquire whether the tendency of Nature to break up certain types into varieties, and form new races—perhaps even new species and genera—was not originally greater than it has been at any period within the knowledge of man. We see no changes take place now, such as long before even the mythic period, produced from one stock the wild urus, the domestic ox and the hunched bull of India. Neither do we see new races of men spring up; such as in the very earliest times produced from one common ancestor the various diverse races of men; white, black, yellow, and red. It is no poetical fancy that Nature's infancy was more active than its later years; that "Nature wantoned in her prime," and produced more gigantic effects than now. Not that the powers of Nature are weakened: but, the purpose having been accomplished, its workings are stayed by the fiat of the Almighty God, and are employed in sustaining and reproducing, instead of in generating anew and creating. When those powers are wanted again, they will spring into undecayed operation; let a new continent rise from the deep and the new world have to be peopled, and Nature will again resume the gigantic forces of its infancy, and become young to fill with life and activity a young world.

But at whatever period impressed, certain it is that many nations have a typical form of Nose, together with other peculiar distinctive features; and it concerns us now rather to regard the fact as it exists than to inquire how it happened.

The Roman, the Greek, and the Syrian forms of Nose have been already descanted upon, as forming three bases of our nomenclature. The present European nations are the Gothic, the Celtic, the Sclavonic and the Finnish.

The Gothic has been subjected to so many varying circumstances that it is now perhaps impossible to assert, with confidence, its original natural form. Where a uniform dull system of despotism, political and religious, has for centuries bound down these nations in abject servitude, the Nose is sharp, devoid of Cogitativeness and Romano-Greek in profile.

This is the case with the Spanish Goths and with those of France and Italy. These nations were so long held in mental thraldom that they ceased to cultivate cogitative powers which it was dangerous to use. Where espionage and *Lettres de cachet*, the Inquisition and Monachism dog and punish men's secret thoughts, and forbid the expression of any sentiment breathing a spirit opposed to the powers that be, or demonstrative of a disposition to inquire into the why and wherefore of political and religious dogmas, the mind, by an instinct of self-preservation, must cease to think. Where to think is a death-warrant, where a look of reflection or an aspect of discontent may be followed by the axe of the executioner, or the more fearful incarceration by the gaoler, the mind has no alternative but to forget itself and live in bestial oblivion, to "sit down to eat and drink and rise up to play." With the cessation of the Cogitative powers, the Cogitativeness of the features will disappear and the Nose will become defective in breadth, thin and sharp. To this want of reflection succeeds, in the naturally higher and more energetic nations, animal passion ; and, if ever the pressure is removed from the national mind and it obtain the upper hand of its keepers, fearful retribution and sanguinary revenge inevitably ensue. They who lived the animal life of a caged wild beast in apparent ease and quietude; well fed and perhaps, sensually, better provided for than if left to their native freedom, will, when let loose from confinement, fearfully vindicate the natural law of liberty, and with an insane instinct tear in pieces the keepers who have fed them for their own purposes and nurtured them for their own pleasure and profit, reckless of the natural social rights of man.

It is for this reason that the sharp, thin unthinking Nose appears symbolic likewise of cruelty; not so much because the natural disposition is cruel, as because the mind, when unchained, acts from animal impulse and not from sage reflection; and animal revenge is always wild and cruel.

We say this of nations which, like the Gothic

and other Caucasian races, were originally well organized and endowed with higher capacities. This higher organization exhibits itself—whatever the degrees of Cogitativeness which incidental circumstances may have added, or adeemed—in a profile, Roman, or Greek, or compounded of both, and which may therefore be called nationally Romano-Greek. The profile not being so subject to variation from the pressure of external circumstances as the breadth, remains still pretty uniformly the same in all the Caucasian races in Europe, which might be written $\frac{\text{I}}{\text{II}}$. Other races there are which, either naturally of less penetrable stuff and a lower and more obtuse organization, or longer ground down beneath a more crushing and uniform despotism, remain contented slaves and willing bondmen. This degradation, as we shall see when we come to speak of the Asiatic nations, appears also in their Noses.

France, Spain, and Italy have been depressed not only beneath a political despotism till within a very recent period, but under the still more soul-crushing despotism of a gross superstition and corrupt religion—the latter even more than the former has repressed Cogitativeness in those nations. If there is one subject which more than another interests the human mind and occupies the thoughts it is its religion—its eternal prospects—for Man is essentially a religious animal. Debarred from exercising thought upon its most natural and interesting topics—and all other subjects being dragged within the jealous circle of a religious despotism—so stern a barrier is opposed to thought that the mind rarely dare overleap it. While a political despotism may be well-pleased to see its subjects occupied in scientific or metaphysical researches, in order to wean them from too critical an examination of itself, a religious despotism forbids any such researches unless made within the small circle it has prescribed. Death or imprisonment awaits a Galileo or a Copernicus, as it would under a similar rule, even now, await a Buckland or a Lyell.

At present, we lament that we can see nothing in the recent revolutionary movements in France and Italy, to indicate the existence of those Cogitative powers, the want of which has always hitherto checked their advancement towards true liberty and self-government.

Now, as in 1793, there seems "equally a want of books and men;" without which, after a few years of bloodshed and anarchy, those countries must again submit to a despotic form of government. No country can be governed without intellect; and if that is not to be found in the many, the few who possess it must become the ruler.

This country has never long needed such a despotism. Germany too, though hardly yet freed from a political despotism, has through a large portion of its area long thrown off the despotism of Rome and embraced the more elevating and life-giving doctrines of the Reformation. In those provinces where this blessed change has taken place, Germany is starting rapidly into that career of intellectuality which England commenced three hundred years ago. The Germans and the English are preeminently deep-thinking nations; and in both of them is the Nose more decidedly and more generally of a Cogitative form than in any other Gothic nations.

The Cogitative may, therefore, perhaps, be said to be one of the characteristic forms of the Noses of those Gothic branches, and might be expressed thus, $\frac{\text{I} + \text{II}}{\text{III}}$. Nevertheless, various degrees of education and various pursuits, with (in England) free institutions, have so diversified their features that they exhibit a much less uniform character than the features of most other nations.

The Anglo-Americans afford a further corroborative proof that the Cogitative Nose is dependent on the cultivation of a Cogitative mind. They present a striking contrast to their puritan forefathers,—men who abandoned home, country and friends, for the sake of religious and political opinions; men to whom conscience was dearer than life, and freedom more precious than worldly advantage; men of the strictest integrity, the most scrupulous honesty, and the sternest firmness, sullied only by an excess of over-wrought feeling—fanaticism. All these virtues, and this vice (itself a virtue gone mad), are wanting to the American character. That there are happy exceptions, it is true; but a nation which boasts *smartness* as its most prominent virtue, must not complain if it is accused of want of principle. The circumstances of young America have contributed to render her's an unthinking people. The wild life to which so large a portion have been subjected, cut off from all neighbourhood, debarred from communication with cultivated minds, thrown entirely on the active business of the day for mental food, they have necessarily degenerated from the thinking men to whom they are indebted for their origin.

So far from the American Nose inheriting the Cogitative form of their ancestors', it is thin and sharp; and, as a national nose, the most unthinking of any of the Gothic stock. America is, however, a fast-growing nation; it has had no infancy, but started at once into life, a full-grown youth. There is hope, therefore—of which already some assurance has been given—that it will yet furnish its quota of thinkers to the history of the human mind.

The Celtic races call for less extended observation. As an un-Gothicized nation, the Irish is the only remnant of a people which probably was at one time thinly spread over the whole of Europe. Nearly related to, if not originally identical with, the Goths, yet naturally of a less vigorous constitution, and lower habit of mind, the Celts rapidly gave way before, or irretrievably amalgamated themselves with, their Pelasgian invaders in Greece and Italy, and their Gothic invaders in Trans-Alpine Europe.

Thus, at one time losing themselves in the overwhelming flood of their invaders, like the waters of a lake inundated by the sea; at another, retreating westerly before the on-coming torrent, the Celtic nations have gradually almost disappeared from continental Europe, and alone find a miserable home and wretched abiding place on the most eastern shores of the Atlantic, and the most western corners of the Old World.

If the Atlantic could have afforded them a footing upon its turbulent waters, they would long since have been driven into it by their rapacious invaders. The complaint of the unhappy Celts has ever since they were hunted to the extreme west of Europe, been the same, "Our enemies drive us into the sea; and the sea drives us back upon our enemies."

Saxon ingenuity has, however, at last endeavoured to circumvent the sea. If it cannot receive *into* its bosom the last wreck of the Celts, it can carry them *upon* it to lands still further west, there to pine, and dwindle away and die; out of sight, and therefore out of the mind, of the haughty invader, who turns with well-feigned horror and disgust from the ruin and degradation which he has wrought.

To make room for himself, he expatriates the ancient owners of the soil, not only without remorse or compunction, but with much self-laudation and pharisaical pride, that he has not extirpated them, and has not—only because he could not—adopted the ingenious idea of temporarily sinking an island to purify it for his own undisputed use and enjoyment.*

* "In 1846, which was a year of larger emigration than any that preceded, it amounted to 129,851. But in the first three quarters of the last year (1847), the

Naturally, however, the Celtic is not a low-class race. It may not have been originally so

---

emigration extended to no less than 240,732 persons, almost the whole of them being Irish emigrants to North America. It is scarcely necessary to observe, that history records no single transportation at all to compare with this. The migrations of classical antiquity were only the slow oozings of infant tribes from one thinly-peopled district into another rather less peopled, or rather more fertile. In actual figures, the irruptions from the north into southern Europe were never at one time more immense."

The Government only refrained from assisting this tremendous emigration at the urgent demand of the land-owners, because it was going on as fast as possible without its aid. Bad legislation had driven the Celt to the ocean, and Saxon ingenuity had furnished him a boat to cross it. Famine and pestilence were at his heels. It was unnecessary to do more. What drowning wretch will not catch at a straw? What patient ideot not fly from misery and death? Yet how monstrous to call such flight—" the headlong terror of a panic-stricken army"—spontaneous!

" It was the *unavoidable* misfortune of this emigration to be *entirely spontaneous.* The cry was —'*Sauve qui peut!*' To send out more emigrants at the public expense, or to promise assistance to all who should highly organized, or so mentally gifted as the Gothic: but in its infancy it had virtues which long thraldom has exterminated.

---

emigrate, would only have been adding fuel to the fire, or like attempting to expedite the movement of a crowd locked in a narrow passage, by applying fresh numbers and pressure to its rear. A miserable necessity dictated that as a general rule, emigration should be allowed to retain its spontaneous, unassisted character. * * * The fever, *it is a painful satisfaction to reflect,* raged with equal force in all the British vessels, whether well or ill-provisioned and appointed. Fearful, too, as the loss of life was, both at sea and on landing, *it was not greater than was reasonably to be expected* from the mortality which prevailed, under circumstances rather less unfavourable for health, in the workhouses and other accumulations of Irish at home."—*Times, Jan.* 1848.

History, in its blackest pages records nothing more horrible than the miseries of the passage; yet while we are maudlin over the horrors of the slave-trade, we 'reflect, with a painful satisfaction,' on the more dreadful sufferings of our fellow-citizens. The slave-dealer calculates to land at their destination four-fifths of his cargo; and it is thought sufficiently shocking that 1 in 5 die on the passage. But the mortality on board the Irish emigrant ships was greater. Many vessels, from their rotten state, perished altogether, with from

It is no fiction that Cæsar found the Gaulish and British youth more apt than the Goths, at acquiring the arts and language of Rome, and that, in a few years, Roman civilization, more efficient than Saxon, had converted Britain into one of the most fertile and well-ordered provinces of the empire. It is no fiction that

---

200 to 300 passengers. This rarely happens with a slaver as the vessels are necessarily of the very best construction. But, of those who escaped shipwreck, 1 in 3, and 1 in 4 died on the passage from fever, and one half the remainder suffered from disease. The 'Laren' from Sligo sailed with 440 passengers—108 died and 150 were sick. The 'Virginius' sailed with 496 passengers—158 died, 186 were sick, and the remainder landed feeble and tottering. It could hardly be otherwise, when vessels built to pack 200 emigrants sailed with twice that number: so that they are described to be worse than the black-hole of Calcutta. And this was the emigration which the British parliament—which laboured to put down the slave-trade—declared itself willing to encourage, had it been necessary from any backwardness in the wretched Celts, to avail themselves of it, and which a British Minister coolly declared it would have been inhuman and unjust to interfere with.

Rome found in Britain one of the most determined opposers of its claim to universal dominion, and that if it were to be

> "Asked, why from Britain Cæsar did retreat?
> Cæsar himself might whisper—he was beat."

It is no fiction that after British Christianity had been driven by Saxon Paganism from Britain into Ireland, the Irish Celts furnished the best schools for literature, and the ablest scholars in Europe—and it is no fiction that ever since the Saxon has set foot in Ireland it has continued to droop and decay, until it is now a foul bog of iniquity; a wretched irreclaimable sink of inhuman vice and monstrous infamy.

This is but the caged wild beast gnawing at its chain, and snapping at its keeper, whether his hand approaches to feed or to heal it. Its Cogitativeness has been repressed till it cannot reflect nor appreciate any but physical modes of escape from thraldom.

It may be said, escape lies open to it in self-elevation, in moral rectitude, and industrious exertion; but it is too late for it to see and understand that. We might as well say to the broken leg, walk now, you walked once; or to the encaged madman, calm yourself and be free, you were calm and free once.

We need no better proof of the non-cogitativeness of Ireland than the facile manner in which it throws itself beneath the Jugernautic car of every demagogue, and sacrifices itself to his avaricious cruelty. We need no better proof of the truth of our theory, than that the Nose of the same nation is deficient in Cogitativeness, and is, for the most part thin and sharp. It has, not, however, lost the Romano-Greek profile, usual among the Caucasian races.

The lowest organized race of any consequence in Europe, is the Sclavonic.

The Sclavones came into contact with Roman civilization earlier than the Goths; but, unlike the latter, they retired to their settlements without carrying away any portion of the manners and habits of the people whom they invaded. Even yet they are but little advanced, since that early epoch. At least till within the present century, the Russian noble, as well as his serf, led the life of a pig, eating and drinking and sleeping. Wallowing in filth, insensate with brandy, and degraded by lust, the Russians of various ranks, differ only in the size and splendour of their respective styes. To enter with minuteness into the daily habits of all classes of both sexes would be to present a picture which we should revolt from drawing, and the reader from beholding.

The Snubby-Celestial form of the Sclavonic Nose, stamps its character irretrievably, and accords remarkably with the description of the Sclavonic mind given by Kohl and other recent writers:—"Inconsistent and unstable—wanting in the creative faculty; but we cannot deny them a marvellous aptitude for all kinds of work, and an extreme facility of imitation."* This is just the description a farmer might give of his horse, or a fine lady of her monkey. "The hope of Europe," says the same author, "from Russian power consists in its total want of vigorous characters, mighty minds, and moral energy." The pictures, which the lively writer Kohl gives of the Russians — their ' small

* Schnitzler's Russia under Alexander and Nicholas.

shrewdness and fox-like common sense', their impudent acknowledgment of their shameless cheating and pedlaring dishonesty—accord literally with the indications which we have ascribed to the Celestial Nose; but we must refer the reader to his work on Russia for endless confirmations of our assertion.

Russia may rise above its present animal degradation, but it will never take a high place in the history of civilization. It may be doubted whether it will ever take any station there at all, except when in some future and long distant age, it is recorded, that, like Asia and Africa, Europe fell from its palmy state, and became a heap of ruins before the furious desolation of barbarous swarms from the north.

Napoleon said, with the prophetic vision of old experience, for

> "Old experience doth attain
> To somewhat of prophetic strain,"

that in fifty years Europe would be Republican or Cossack. He only erred in using the disjunctive; for it does not require much penetration to foresee that, at no very distant period, Europe will be both—first Republican, and then, when thus prostrated at the foot of the first powerful despot—Cossack.

For this purpose, it is probable the Sclavonian nations, with hordes of Mongolian Calmucks, and Tartars—the σιμοὶ, or flat-nosed nations, of Herodotus—are gathering force and increasing in their vast plains and desolate forests. The scourge of Europe is being prepared; slowly but surely; and when civilization shall have taken a firm hold of America and the new continents gradually being built up in the Pacific, Europe, having fulfilled its part in the world's history, will be swept away, and become a byword and a scorn among the nations—Ichabod will be written on its temples, and the bittern and the owl shall inhabit it; the wild beast of the desert shall lie there, and the dragons in its pleasant palaces.

The Finnish race presents a remarkable proof of the variation in physiognomy attendant on variation in mental capacity, occasioned by change of circumstances—as government, climate and habits. The ancient Huns, the modern Hungarians, and the northern Finns and Lapps of the shores of the Bothnian Gulf and the White Sea, are all of the same race; and yet differ widely from each other in physiognomy and psychonomy.

The differences between those races took place within the historic period, and afford a striking instance of the effect of external circumstances in modifying the mental and corporeal features.

The fierce and savage Huns, who overrun a portion of the Roman Empire under Attila in the fifth century, differed wholly from the Finns now existing in Europe. So misshapen were their features, and so hideous their aspect, so savage and demoniacal their warfare, that the terrified Goths could not believe them to be born of woman, but asserted them to be the unnatural offspring of demons and witches in the fearful solitudes of the icy north. One of their distinctive features was a flat depressed Nose, plainly indicating their low organization.

Although the Finns and Lapps retain the flat-nose—never having emerged from barbarism—they are a mild, gentle, meek-spirited race, presenting few features which seem capable of amelioration.

The Hungarians, on the other hand—in whom, however, we must suspect a large infusion of Gothic blood—are a bold, independent, noble-minded, and highly intellectual people; characteristics which exhibit themselves in a noble Roman Nose, and a countenance bespeaking the independence of their minds.

We may next advert to the characteristic features of a few of the Asiatic nations.

Perhaps no nation displays a more universal dead level and general sameness of feature than the Snub-nosed Chinese. Notwithstanding the great varieties in climate and soil which prevail in that extensive Empire, and the correspondent variations which must be made in domestic habits and style of living, a remarkable identity of feature prevails among all classes of every province. The faces may be said to be all cast in the same mould, and one could wish that Nature, when she made the first cast, had—as she is reported to have done when she made a certain beautiful female, whose name we forget—broken the mould before she produced any more casts from it. Perhaps, however, we belie the good old dame in attributing the production of this, or any other equally ugly countenance to her. It is rather the degraded form into which a despotism of unknown duration and unexampled soul-depressive powers has converted the original type.

A form of government more admirably arranged to keep the people in a state of childhood has never been modelled than that of China. The wisdom of its arrangements for securing the permanent despotism of the ruler is undeniably proved by its long and peaceable subsistence. To rebel in China is the heinous crime of filial disobedience: it is not, as in Europe, a political crime merely, it is also a moral crime of the same class as murder or theft. Unless we can imagine a nation by universal assent throwing off the bonds of morality, and living in confessedly gross crime, we can form no conception of Chinese rebelling. It would present the unnatural and inconceivable state of a nation of parricides and disobedient children.

Every superior in China, from the Emperor

to the military officer or civil Mandarin, is "a father;" all under him are his "children," and as such must obey him without question or demur. "Filial disobedience," whether to parents or governors is the highest crime. Filial disobedience is thus defined:—"In our general conduct not to be orderly, is to fail in filial duty: in a Magistrate not to be faithful, is to fail in filial duty: among friends, not to be sincere, is to fail in filial duty: in arms and war, not to be brave, is to fail in filial duty." A people thus treated as children, must ever remain in a state of childhood; and though education is general among the Chinese, it is an education which, like the bandages on their women's feet, binds their minds *from* growing, and restricts them to the size and calibre of infancy.

Education in China consists solely in social and political training for the purposes of despotism. The studies are confined to one unvaried routine, and no deviation from the prescribed track is permitted. Within this circle all are, and must be, educated. Hence an uniformity of mind prevails, and has prevailed for ages throughout China, and has extended itself to the national features; betraying itself in Snub Noses and a dull, stolid expression of countenance. So much for compulsory education! It is impossible that it should be otherwise. A nation whose minds are all reduced to the same level; whose thoughts are prescribed; whose daily conduct is measured out; whose very amusements are dictated by an imperial will, must necessarily soon become uniform, both mentally and physically.

This uniformity will be the waveless level of the Dead Sea. Storms may agitate the upper sky, winds may rage, and floods descend; but the waves are too heavy to rise from their death-like repose. They sleep the calm sleep, not of peace, but of death. The last trumpet alone can arouse their torpor. The benignant mind of the Christian may nourish sweet hopes of evangelizing a nation so sunk, but the hopes are vain. Christianity came not till the human mind was fitted and prepared to receive and understand its divine precepts. It came not to the infancy of the world, but to its old age and matured judgment. A nation, therefore, steeped in the irreclaimable dotage of a childhood which has endured throughout its whole life, cannot receive it. Both the Hindoos and the Chinese have forfeited by their long-lived puerility the blessed message.

The first and every subsequent step of Christianity, as of civilization, has been Westward. Neither can ever return to the East. The Apostle of the Gentiles preached from Judea to Pamphylia and Galatia, but was forbidden of the Holy Ghost to preach the Word in Asia;* and when he assayed to go Eastward into Bithynia, the spirit suffered him not, but compelled him westerly into Macedonia. From Macedonia to Rome; from Rome to Gaul; from Gaul to Britain; from Britain to America and Polynesia, the course has still ever been uniformly westward.† A few isolated Christians may be made in Asia; but it will never be christianized. Asia has performed its part on the world's stage. It is dead out, and can-

* Acts, XVI.

† That is, westerly from the country last civilized or christianized.

not be resuscitated. When Christianity is entertained by "all nations," Asia will be no more. It will not be reckonable among the nations, even as a dead man is not among the living. This may seem a harsh judgment. But is it harsher that nations whose own degradation unfits them for Christianity, shall remain ignorant of it during the brief remainder of the world, than that they have been ignorant of it for nearly two thousand years?

It is not for man to judge God, and to say that His ways are unjust. We must not deny the fact because we cannot comprehend it. We cannot tell by what crimes Asia has forfeited her part in the New Covenant of Grace. It may be because she rejected the first dispensation and flagrantly violated the Old Covenant of Works. To Asia, the mother of mankind, the blissful seat of our first parents, the nurse of the renovated human race, were given the first pure, simple precepts by which Man was taught to obey his God as a child obeys his parent. How soon she flung off this obedience and rejected her Great Teacher let history, both sacred and profane, attest. Long ere Asia sent forth peoples and nations to replenish other quarters of the earth, these original precepts had been obscured and obliterated by idolatry and polytheism. A lesser crime, therefore, attached to these misinstructed offsprings than to the misteaching mother. A second dispensation was therefore revealed to them, but forbidden to her. So far man might think he comprehended the divine purposes without impugning God's wisdom and justice; yet may he err, and his frail musings be but the cogitations of the flea which reasons on the movements of the elephant, whose back is his universe. This should be the humble reflection of all who strive to justify the ways of God to man. We know but in part, and we see but in part, and therefore cannot judge of Him who sees and knows the whole.

We have incidentally mentioned the Hindoos as partaking in the mental degradation of the Chinese. But, nevertheless, they are not nearly so degraded a race, nor have they so general an uniformity in their features, nor so low a formation of their Noses. India has been subjected to less uniformity of despotism than China.

While to the dominant system of the latter we can assign no limit, we find in that of the former numerous epochs when important changes have taken place.

Fierce religious wars, frequent foreign invasions, domestic feuds and intestine warfare have kept the Hindoo mind more on the alert than that of China. Assyria, Egypt, Scythia, Greece, Persia and Britain have at different epochs overwhelmed India. Idolatrous Monotheism, Polytheism, Mahometanism and Christianity have, in turn, violated its shrines and endeavoured to overwhelm both Buddha and Brahma. Buddha and Brahma, Vishnu and Siva have striven to overthrow each other; but while the country has been desolated, the people have been saved from sinking into the uniform degradation of the Chinese. Nevertheless, under each and every system, despotism has prevailed in India; no free institution has ever flourished on its plains; and, therefore, despite the stirring events which have excited it, it has never risen again to that high station which its people must have held among their cotemporaries when they sculptured the caves of Elephanta and Ellora, and raised the pyramidal pagodas of Tanjore and Deogur.

These gigantic works sufficiently attest that the inhabitants of India are not naturally of a low-class race. Forty thousand men labouring incessantly for forty years would hardly suffice to excavate and sculpture the cavern-temples of Salsette alone. Yet those form but a small portion of similar gigantic works of the same age.

No mean-minded men raised fanes such as these to the Deity. Energy of the most vigorous character, talent of the highest rank, and devotion of the noblest nature could alone have dictated and executed structures which outvie in magnitude the boldest efforts of modern genius. In comparing them with the latter, we should moreover recollect that they were the first efforts of the human race; made without pattern, designed without exemplar, and commenced and carried out without experience.

How different must those men have been from the soft and effeminate Hindoo who has forgotten in the mist of ages these shrines of his fathers, and abandoned them to ruin and decay; and who, conscious of his own utter inability to achieve or conceive their equals, ascribes their formation to giants and demigods. And they were different. The same race, but different men, different in features as in minds. While the profile of the modern Hindoo is soft and effeminate, and the Nose short and rounded, (Parabolic) the ancient sculptures demonstrate that the profile of their earliest progenitors was manly and decided, and identical with that of their descendants, the Indo-Germanic nations, in Europe. One well-known instance will suffice. The Trimurti or three-headed deity in the caves of Elephanta.

THE HINDOO TRIMURTI.

This is a sculpture of the most remote antiquity, but the dress, the beads, the sacred cord and other religious symbols declare it to be the work of Hindoos. In anthropomorphising the Deity, men always adopt their own typical countenance for that of their God. Hence their idols betray the National features. Now, observe the profiles of Vishnu and Siva in this Trimurti. The face of the former, the good and beneficent "Preserver," the friend and mediator for Man is a purely Greek face; the Nose straight and well-defined. It has none of the air of the modern Hindoo countenance. Much less has that of the energetic and terrible Siva, "the Destroyer." The Nose is of the most energetic form; it is a fine Roman Nose, aquiline and rugose. If phrenologists are permitted from similar facts to say that the Greeks—who were but children to these Hindoo artists—were phrenologists, surely we may venture to say that even at this very early period the Hindoos were Nasologists.

But in the wide nostril of Brahma we also perceive the Cogitative form of Nose, so necessary to indicate the wisdom of Brahma, "the Creator:" who, though he now rests, having consigned the inferior office of Preservation to Vishnu, was the first emanation from the supreme Brahma, and by whom and from whom all creation proceeded. With the exception of the head in this Trimurti, Brahma has no idolatrous representations, for it is said in the Vedas, "Of Him whose glory is so great, there is no image. He is the incomprehensible Being which illumines all, delights all, and whence all proceed."

Sir William Jones mentions in one of his discourses published in the Asiatic Researches, the existence of a small nation in India which appears distinct from the Hindoo race. The people comprising it he describes as shrewd, clever tradesmen, enterprising merchants, acute money-lenders, and notorious in India for their aptitude for commerce. Their countenances, he adds, are what are called Jewish, and hence he concludes that they constitute a portion of Jews, who either at the dispersion of the Ten Tribes, or at some other very early period settled in

India. It is surprising that the acute President should have so hastily jumped to such a conclusion from the foregoing premises; for he adds a fact which seems most decidedly to negative it. This people, he tells us, have not the slightest trace of any Jewish traditions, belief, or customs among them. Now it is a familiar fact that the Jews, wherever dispersed, or however long separated from their brethren, have invariably retained a very large proportion of the inspired precepts revealed to regulate their religious, moral, and social conduct; and it must demand the most precise and indisputable evidence to justify the classing any people as Jews, who have lost all traces of the manners and customs of that singular nation.

For these reasons we do not hesitate to say that the two facts on which Sir W. Jones founded his hasty hypothesis, viz., the commercial character and the Jewish physiognomy of this Asiatic tribe, afford by their coincidence only a remarkable and curious confirmation of our Nasological theory, and as such, we here gladly insert it.

We have said that the Jewish Nose should more properly be called the Syrian nose; but have reserved until this place, some of the corroborative illustrations.

The Syrian Arabs, as descendants of Abraham, through the wild son of Hagar, inherit the physical, and many of the metaphysical, features of the Hebrew nation.

Destined by the promise of God to become a great nation, the Arabs founded one of the most extensive kingdoms of the earth, and for many centuries swayed an empire more extensive than that of Rome in her fullest prosperity. For twelve hundred years, a larger proportion of the inhabitants of the earth have devoutly obeyed the precepts of the Arabian prophet, than have knelt at the altar of any other individual creed; and, though Mahometanism is perhaps doomed to fall before Christianity, it cannot be regarded in any other light than as a minor dispensation, and an inferior blessing conferred by Providence on a very large portion of His people.

Christians, who yet recognize the finger of God in every sublunary affair, would shrink with horror, if asked to recognize in Mahometanism a Providential dispensation; yet, whether we regard it as a religion which annihilated the grossest idolatries, abolished human sacrifices, exterminated the vilest obscenities, and substituted a nearly spiritual worship of One God, over the largest and fairest portion of the earth,—or as the religion of a nation, whose ancestor God blessed, and promised to "make a great nation," and "to multiply exceedingly, that it should not be numbered for multitude;" and who, in token thereof, received the seal of circumcision—to this day retained as among the Jews—it is difficult not to see in it the finger of God, or to deny that the pseudo-prophet of the sons of Ishmael was an unconscious instrument in His hands.

But this is a topic not needful for us here to enter fully upon. It is more to our purpose to remark upon the psychonomic features of the Arabs, while in the zenith of their glory as a nation; when the Caliphs of the East ruled as Priest and Potentate, over more than two-thirds of the known globe.

During this glorious period of their power, the Arab character shone out uncontrolled in its true features, and exhibited itself as it had never done before, nor since.

True to its parentage, but unshackled by the stringent laws, and anti-social ceremonies of its more favoured brother, it rioted in all those tastes and pursuits which the latter delighted in, but was restrained from; and became celebrated for a splendour, which was rivalled by that of Solomon alone, and a traffic which far outvied that of all contemporaries, or predecessors — except, perhaps, the cognate nation, the Phœnicians.

Rich in barbaric pearls and gold, and boasting all the wealth of Ormuz and of Ind, the Court of the Caliphs verified the visions of the "Arabian Nights;" which, if true, were true here only. All the gauds and trinkets, the golden palaces, the jewelled walls, the glittering roofs, in which the other branch* of the Hebrew nation displayed their highest ideas of magnificence, shone resplendent in the Halls of the Caliphs.

But as to the boasted literature of the Arabs, it resolves itself into an ardent pursuit of physical science—astronomy, chemistry, and the mechanical arts, for nearly all the more important of which we are indebted to the Arabs; not, however, as inventors, but as carriers, like the Phœnicians. In the higher departments of literature, the Arabs made no progress. Metaphysical disquisitions, and intellectual pursuits were repugnant to their tastes, which rather delighted in the physics of Aristotle than the metaphysics of Plato.

Nor were they less true to their nasal development in their success and skill in com-

* The Hebrews consider themselves to be so named from Heber, an ancestor of Abraham (Gen. xi. 15). The descendants of Ishmael are therefore equally entitled to the name.

mercial pursuits. The commerce of Arabia, for several centuries, encircled the whole known world. From the frigid shores of Scandinavia, from the torrid sands of Africa, from silken Cathay, from jewelled Ceylon, from vine-clad Europe, from spicy Araby, flowed the rich streams of produce. The amber of the north was exchanged for the gold of the south; the wines of Spain for the silks of China; the pearls of Ceylon for the slaves and gold dust of Africa; and a commerce now excelled only by that of England, carried arts and literature from one end of the Old World to the other, and was mainly instrumental in raising the more highly organized nations of Europe from barbarism to a physical and intellectual splendour hitherto unknown.

But from this glorious reality, the Arab has sunk into a wretched, irretrievable lethargy. Like the Jew, he has been weighed in the balance and found wanting; the cup of promise has been held to his lips, and he has refused, or polluted the blessed draught. They have been called, but would not come, they would have been gathered together as tender chickens under the wings of the hen, but they would not; and "behold their house is left unto them desolate."

Neither Arab nor Jew shall ever again revive, till they join with the whole earth in one universal cry, "Blessed is He that cometh in the name of the Lord!"

It has been said that Christian intolerance has driven the Jew into the mart, and sunk his soul in barter. But this is not true—Commerce and money-getting are the psychonomic features of both the Hebrew races. The Israelitish branch is vehemently charged with its usury and extortion, by all its prophets. The severe laws which Moses made against usury shew the character of the people for whom they were necessary; yet those laws were ineffectual to check this inherent vice. Ezekiel (cap. xxii. 12) exclaims, "Thou hast taken usury and increase, and thou hast greedily gained of thy neighbours by extortion, and hast forgotten me, saith the Lord God;" so all the prophets.

The Arab and the Jew are both now equally sunk in the same degradation, (*Heu! quantum mutati!*) and both exhibit, through this degradation, their love of gold, though in a different manner. The Arab still haunting his native soil, from which legitimate commerce is almost excluded, betrays his ruling passion in extortion from travellers, in skilful chicanery in horse dealing—the only commerce left to him—or in impudent incessant demands on strangers for *bacsheesh.*

All travellers agree, that when the Arab, degraded as he is, has an opportunity, there is no shrewder or more skilful bargain maker, nor any one more competent to extract by ingenious chaffering, the full equivalent for his services. He has been designated by fleeced and angry travellers—little thinking how near the mark they were—the Jew of the desert. The Jew, driven from the land of his birth into a wider sphere, turns his commercial propensities to better account, and under every clime, and amidst every race, out-manœuvres and surpasses his less shrewd antagonist.

Other Asiatic nations might seem to call for observation; but so little is known of their mental characteristics, that it would be improper to endeavour to substantiate our cause by them.

It is unnecessary to do more than remind the reader of the low development of the Negro mind and his miserable nasal conformation—they are worthy of each other. However humane may be the attempts to elevate the Negro, it never can be done till his Nose is more elongated; but as its present form has subsisted without alteration for three or four thousand years, there does not seem much hope of its being improved now. The Negro race, as old as the earliest Egyptian sculptures, has never risen to an equality with any of the other races; and, though we would not willingly condemn any nation to hopeless degradation, yet the history of the Past *will* reveal somewhat of the secrets of the Future, and he is a fool who cannot, and a coward who dare not, read them.

As among individuals so among nations,

there are orders and degrees of mind, and it is only the blind who cannot see that the equality of the one is as wild a dream as the equality of the other.

In the new Islands of the Pacific, we behold a constant succession of new worlds emerging from the deep by means of the same process which, in the pre-Adamite world, formed and elevated the islands and continents of the Northern hemisphere. Minute polypi are secreting from the waters, and fixing on the summits of submarine volcanoes, the solid and durable limestone which now forms their protection from the waves, and which will hereafter form the foundations on which accumulated detritus will heap up fertile soils and habitable lands.* Earthquakes are continually

* "The prodigious extent of the combined and unintermitting labours of these little world-architects must be witnessed in order to be adequately conceived or realized. They have built up 400 miles of barrier reef on the shores of Caledonia; and on the north-east coast of Australia their labours extend for 1000 miles in

pushing up these horizontal surfaces, and breaking them up into mountains which, arresting the clouds in their progress, draw down into the valleys and plains the fertilizing rain. This smooths down the asperities of the earthquake-broken surface, and softens and harmonizes it into that sweet variety which gives birth to

"The pleasure situate in hill and dale."

To people these new lands, Nature has branched off from the old stock, new races of men of various degrees of physical development and intellectual endowments. While those nearest the old continent of Asia, and there-

length; averaging a quarter of a mile in breadth, and one hundred and fifty feet in depth. The geologist, in contemplating these stupendous operations, learns to appreciate the circumstances by which were deposited, in ancient times, those mountain masses of limestone, for the most part coralline, which abound in many parts of our native island."—*Ansted's Ancient World*, p. 32.

fore nearest to the old blood, are of the lowest possible mental and physical organization, little elevated above the low-class animals—the kangaroo and the ornithorynchus*—of the Australian plains, those at a greater distance—the New Zealander and the Otaheitan—exhibit a development which may vie with that of the Caucasian nations; and which has proved its equality by not sinking before them, but maintaining against Saxon invaders equal rights and equal privileges.

We have a striking instance of this before us at the present time. The British Legis-

---

* Zoologists class the Marsupiala as the very lowest form of Mammalia, and but little removed above the cold-blooded Reptilia. They are a connecting link between those two great classes of Vertebrata. The Ornithorynchus is an animal of still lower organization. The whole fauna and flora of Australia indicate a newly-formed land, and are analogous to those of the Poilitic and New Red Sandstone Ages of the Northern Hemisphere; which in like manner succeeded Coralline Limestones, and in which small islands began to be united into large islands and quasi-continents.

lature having, in ignorance of the determined character and clever good sense of the New Zealanders, endeavoured to force upon them a Constitution which deprived them of legislative privileges equal with those of the colonists, and which gave to the latter the power of taxing the former without their consent, the natives have resented the injustice so firmly, but hitherto peaceably, that the Governor, Sir George Grey, has been compelled to suspend this so-called Constitution, lest it should foment a war of the most deadly character. It is worthy of observation that the injustice attempted to be done to this shrewd and spirited people, is not one of an evident physical character, such as any savage can appreciate, but one of a purely theoretical and political nature, the importance of which is even yet hardly sufficiently understood and appreciated in any country besides England. Sir George Grey writes to the Home Government as follows :—

" By the introduction of the proposed constitution into the provinces of New Zealand, her Majesty's Ministers would not confer, as it was intended, upon her subjects the blessings of self-government, but would be giving power to a small minority (the colonists). She would not be giving to her subjects the right to manage their affairs as they might think proper, but would be giving to a small *minority a power to raise taxes from the great majority* (the Aborigines). There was no reason to think that the majority of the aboriginal inhabitants would be satisfied with the rule of the minority; while there were many reasons for believing that they would resist to the uttermost. They were a people of *strong natural sense and ability*, but by nature jealous and suspicious. Many of them were owners of vessels, horses, and cattle, and had considerable sums of money at their disposal, *and there was no people he was acquainted with less likely to sit down quietly under what they might regard as an injustice.*"

"For these and other reasons, the Governor announced that he should not proclaim the constitution before receiving fresh instructions from the Colonial Office.

"The tone of the most trustworthy correspondence from New Zealand, proves that this exercise of independent authority on the part of Governor Grey has saved the colony from disastrous consequences. Ministers acknowledge his superior competency to judge in a matter of this kind, and a bill has accordingly been introduced into the House of Commons by Mr. Labouchere, 'for suspending, during a limited time (viz. for five years), the operation of part of the act for making further provision for the government of the New Zealand Islands.' "*

Thus has this noble people, with a strong natural sense and ability not hitherto supposed to belong innately to "savages" opposed more successfully the first step in tyranny—the power of unrepresented taxation—than any other nation (except the Saxon), which has ever existed, civilized or uncivilized.

This has been done within twenty years after their actual beneficial contact with civilization;

* Leeds Mercury, Jan. 1848.

but it was more than 600 years after the Norman conquest, before the Saxon roused himself to enforce the same right of self-taxation. There could be but one better evidence than this of the high class mind of this people; and it has furnished this one better, and best evidence—its speedy and conscientious reception of Christianity.

While for several centuries missionaries of every denomination have laboured in Asia in vain; no sooner was Christianity efficiently made known to the New Zealanders, than catching at once with a remarkable aptitude its leading characteristics, and, appreciating immediately its beneficent doctrines, they accepted it; and now, together with other Polynesian islands, New Zealand affords the proudest conquest and the richest harvest of the soldier of Christ.

Yet apparently, for no nation could Christianity be less adapted, and no nation could be expected to afford less hope of speedy conversion. The Pagan New Zealander was a fierce, blood-thirsty monster, spending his whole life, and finding all his pleasures in the most savage warfare. Not content with slaying his enemies in combat, he sat down afterwards with a joyous enthusiasm worthy of a fiend, to make a feast on their carcasses. Human sacrifices stained his altars, and hideously-deformed images pourtrayed his debased notions of a God.

On the other hand, the peaceable and mild Hindoos, whose religion forbids bloody sacrifices of any kind, and enjoins the careful preservation of the spirit of life, even in the meanest forms; whose singular traditions of the incarnate Chreeshna seem to point distinctly to a Messiah, and whose remarkable Trimurti, three in one, and one in three, seems to open a way to the facile reception of the difficult doctrine of a Trinity in Unity, have never, as a nation, a province, or even a small village, embraced Christianity. China, which has its similar traditions, whose sages have taught that "The true Holy One is to be found in the West," and that "Eternal reason (Λογος) produced One, One produced Two, Two produced Three, and THREE produced all things," and whose calm stoicism and severe morality are so accordant with the external symptoms of a Christian mind, has hardly furnished a single convert, and apparently feels no curiosity about the religion of the Fanqui (white devils).

If history is the past teaching lessons to the future, surely our Missionary Societies might take a lesson from these facts, and withdraw their exertions from so hopeless a field as Asia, and expend them on the hopeful soil of Polynesia. Surely if the great Apostle of the Gentiles, who was specially appointed to bring into the fold of Christ "all nations" was forbidden to preach the Word to the effete nations of Asia, it is not given to his successors to contravene the inspired mandate.

Other injunctions of Scripture to the apostolic Church are rightly interpreted as applicable, and to be obeyed by, the Church in all future ages; and it is a strange inconsistency, arising from a too warm and enthusiastic desire to promote the kingdom of Christ, fruitlessly to strive, in this instance, against the mandate of the Holy Spirit.

Thus much we have said, to contrast the

New Zealand mind with the Hindoo and the Chinese, because the same contrast is manifest in their respective physiognomies.

NEW ZEALANDER.

Compare the bold energetic Roman Nose, the manly and commanding profile of the New Zealander, with the soft and rounded features of the Hindoo, and the flat monotonous surface of the Chinese visage. You perceive at a glance that the first is the face of a man of strong straightforward common sense, and intense energy. He may not be an acute and subtle reasoner; but he catches at once the leading points of a subject, instantly decides, and instantly acts upon his decision.

While the two latter remain in imperturbable absorption, and while the subtle "Greek" would be thinking too precisely on the event,

"A thought which, quartered, hath but one part wisdom,
And ever three parts coward,"

the "Roman" has been, and seen, and conquered. He is come back, at home, resting after his successful toil; while the "Snub" is thinking about getting out of bed, and the "Greek" is making up his mind whether it is "worth while" to go out.

Thus we have, from divers sources, brought together, briefly and succinctly, a few of the universal proofs which establish Nasology as a science. From individuals and from nations we have gathered the basis of our nasological laws; and we trust we have produced conviction in some minds that "the Nose is an index to Character;" if not, we shall not say to the reader, as phrenologists do to their incredulous auditors, that it arises from his defective organization, but rather attribute it to our own defective mode of argumentation; for we shall not willingly admit the erroneousness of a system which has been built up upon many years of personal observation both among the dead and among the living.

THE END.

LONDON:
Printed by Schulze and Co., 13, Poland Street.